Cambridge Natural Science Manuals
BIOLOGICAL SERIES.

GENERAL EDITOR :—ARTHUR E. SHIPLEY, M.A.
FELLOW AND TUTOR OF CHRIST'S COLLEGE, CAMBRIDGE.

PRACTICAL
PHYSIOLOGY OF PLANTS.

PRACTICAL
PHYSIOLOGY OF PLANTS

BY

FRANCIS DARWIN, M.A., F.R.S.,

FELLOW OF CHRIST'S COLLEGE, CAMBRIDGE,
AND READER IN BOTANY IN THE UNIVERSITY,

AND

THE LATE E. HAMILTON ACTON, M.A.,

FELLOW AND LECTURER OF ST JOHN'S COLLEGE, CAMBRIDGE.

WITH ILLUSTRATIONS.

SECOND EDITION.

CAMBRIDGE:
AT THE UNIVERSITY PRESS.
1895

[*All Rights reserved.*]

CAMBRIDGE UNIVERSITY PRESS
Cambridge, New York, Melbourne, Madrid, Cape Town,
Singapore, São Paulo, Delhi, Tokyo, Mexico City

Cambridge University Press
The Edinburgh Building, Cambridge CB2 8RU, UK

Published in the United States of America by
Cambridge University Press, New York

www.cambridge.org
Information on this title: www.cambridge.org/9780521230803

© Cambridge University Press 1894

First published 1894
Second edition 1895
First paperback edition 2011

A catalogue record for this publication is available from the British Library

ISBN 978-0-521-23080-3 Paperback

PREFACE.

EDWARD HAMILTON ACTON died in the early part of the present year. This is not the place to speak of what we, his friends, have lost by his death, nor of his promising career as a man of science. All that I can do here is to explain in what way I have dealt with that part of the book for which he was especially responsible.

Part II, on the chemistry of metabolism, was entirely written by Mr Acton, and is here reprinted practically as he left it. The only changes are two or three corrections found in his interleaved copy of the book, and a few verbal and typographical alterations introduced for the sake of uniformity.

The book (as explained in the preface to the 1st Edition) originated in the following way.

In 1883 I began a course of instruction in the physiology of plants, of which the chief feature was the demonstration of experiments in the lecture-room. Some

years later a different arrangement was made ; the students
were required to perform the experiments for themselves,
and at the same time laboratory work in the chemistry of
metabolism was organised by Mr Acton. To enable the
students to carry out their work, written instructions
were needed, and the present book is the result of an
extension and elaboration of what we prepared for our
classes.

The book makes no pretence to completeness, it
contains merely such a selection of experimental and
analytical work as seems suitable for botanical students.

Part I, which deals with general physiology, is
necessarily of a somewhat more elementary character
than Part II, which treats a particular department of
physiology in a more special manner, and presupposes a
greater amount of knowledge on the part of the student.

A few experiments which experience has shown to
be unsuitable have been omitted in the present edition.
The chief additions are :—Exps. 5 and 52 (Timiriazeff's
eudiometer), Exp. 33 (the importance of stomata in
gaseous interchange), Exps. 118 A, 118 B, 118 C (Stahl's
cobalt method), Exp. 205 A (Pfeffer and Czapek's method
of localising geotropic irritability in roots), Exp. 249 A
(chemotaxis in Bacteria), Exps. 249 B and C (chemotaxis
in pollen-tubes)

The references to the literature of Part I have been increased in number, and they now give a fuller, though still but a rough guide to the published authorities. The references to Sachs' books appear to have been a source of difficulty to some of our readers. I therefore give the full titles of those to which reference is made.

Physiologie Végétale, recherches sur les conditions d'existence des plantes, et sur le jeu de leurs organes. Traduit de l'Allemand avec l'autorisation de l'auteur, par Marc Micheli. Paris, V. Masson et Fils, 1868. [This is the translation of Sachs' *Handbuch der Experimental-Physiologie der Pflanzen,* being volume IV of Hofmeister's *Handbuch der Physiologischen Botanik,* Leipzig, 1865.

Arbeiten des botanischen Instituts in Würzburg, herausgegeben von Prof. Dr Julius Sachs. Leipzig, Engelmann. Band I. 1874, Band II. 1882, Band III. 1888.

Text-book of Botany, morphological and physiological, by Julius Sachs, Professor of Botany in the University of Würzburg. Edited with an Appendix by Sidney H. Vines, M.A., D.Sc., F.L.S., Fellow and Lecturer of Christ's College, Cambridge. Second Edition, Oxford, at the Clarendon Press, 1882. [This is a translation of Sachs' *Lehrbuch der Botanik* (Edit. 1874).]

Vorlesungen über Pflanzen-Physiologie von Julius Sachs. Leipzig, Englemann, 1882. [An English trans-

viii PREFACE.

lation was published in 1887 by the Clarendon Press under the title *Lectures on the Physiology of Plants.*]

Gesammelte Abhandlungen über Pflanzen-Physiologie von Julius Sachs. Leipzig, Engelmann. Band I. 1892, Band II. 1893. [This is referred to in the text as Sachs' *Collected Papers.*]

I gladly take this opportunity of expressing my thanks to Mr F. F. Blackman, Demonstrator of Botany in the University, for much valuable help in the arrangement of the experiments in Part I. Also to the Cambridge Scientific Instrument Company for the use of the clichés for Figs. 25 and 26.

FRANCIS DARWIN.

Botanical Laboratory, Cambridge.
September, 1895.

CONTENTS.

PART I.

GENERAL PHYSIOLOGY.

CHAPTER I.

ON SOME OF THE CONDITIONS AFFECTING THE LIFE OF PLANTS.

CHAPTER II.

ASSIMILATION OF CARBON.

CHAPTER III.

FURTHER EXPERIMENTS ON NUTRITION.

CHAPTER IV.

TRANSPIRATION.

CHAPTER V.

PHYSICAL AND MECHANICAL PROPERTIES.

CHAPTER VI.

GROWTH.

CHAPTER VII.

CURVATURES.

CHAPTER VIII.

FURTHER EXPERIMENTS ON MOVEMENT.

xiv   CONTENTS.

PART II.

CHEMISTRY OF METABOLISM.

CHAPTER IX.

INTRODUCTION.   SOLVENTS.   METHODS OF EXTRACTION.
GENERAL NOTES ON APPARATUS AND MANIPULATION.

Introductory.   Preparation of material to be examined.   Preparation
of extracts: non-nitrogenous plastic substances.   Preparation of extracts: nitrogenous plastic substances.   Filtration.   Evaporation of
solutions.   Changes occurring in solutions on keeping   pp. **237—248.**

CHAPTER X.

CHAPTER XI.

CHAPTER XII.

CHAPTER XIII.

DEXTRINS AND SUGARS, GLUCOSES, CANE-SUGAR, MALTOSE, &C.

CHAPTER XIV.

STARCH. CELLULOSE.

CHAPTER XV.

ORGANIC ACIDS AND SALTS.

CHAPTER XVI.

UNORGANISED FERMENTS. (ENZYMES.)

CHAPTER XVII.

GENERAL EXPERIMENTS.

APPENDIX I.

APPENDIX II.

LIST OF ILLUSTRATIONS.

ERRATA.

Page 67, note 1, *for* " Zimmerman" *read* " Zimmermann."

   ,,   70, line 5 from foot, *for* "developement" *read* " development.'

   ,, 102, last line, *for* "fig. 18" *read* "fig. 20".

   ,, 135, note 1, *for* "1889" *read* " 1888."

   ,, 136, line 2, *for* " rasor" *read* " razor."

PART I.

GENERAL PHYSIOLOGY.

CHAPTER I.

SECTION A. *Respiration.* SECTION B. *Temperature—
Poisons—Electricity.*

SECTION A. **Respiration.**

The presence of free oxygen is a necessary condition of
the life of all the higher plants.   This fact will be more
conveniently demonstrated in the chapters on growth and
growth-curvatures.   The present section is intended as an
introduction to the study of the facts without special
reference to the importance of respiration.

(1) *Production of CO_2.*

Take a stoppered jar of about 500 c.c. capacity, fill it
to one-third of its height with (in spring) horse-chestnut
buds or (in winter) with beans which have been soaked
in water for 12 hours and have been afterwards placed in
damp cocoa-fibre for 12 hours.   Place the jar in a warm
room, and after 12—24 hours cautiously open the jar and
lower a lighted taper which will be extinguished as it
enters the CO_2 produced.

(2) *Absorption of CO_2 by Potash.*

Take a filtering flask of 400 or 500 c.c. capacity, having a lateral opening as shown in fig. 1, to which a

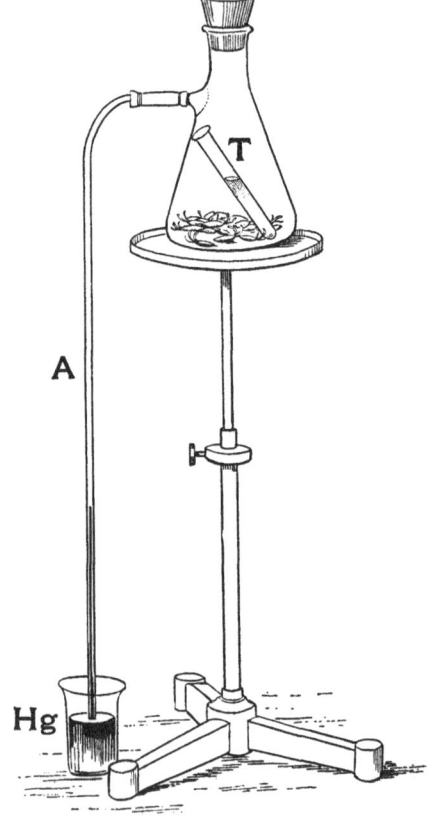

Fig. 1.   Exp. 2.

glass tube, *A* (4 or 5 mm. bore), is attached by thick rubber tubing and wire ties. The end of *A* dips into the mercury[1] in the beaker *Hg*. The flask contains enough germinating barley to cover a piece of wet filter-paper at the bottom of the flask. Barley germinates well in winter : it should be soaked in water for 24 hours and kept in damp air for 24 hours before use. A test-tube *T* half full of strong KHO is introduced into the flask, which is then closed by a sound tightly fitting rubber cork. As the CO_2, produced by respiration, is absorbed by the KHO, the mercury in the beaker *Hg* is sucked up the tube *A*. In starting the experiment it is necessary to warm the air in the flask before the end of *A* is forced into the mercury, so that as the air cools again the mercury may be sucked a little way up the tube to a point which will then serve as zero for subsequent observations. The warming may be done by immersing the flask in water at 40° for a few minutes; or it may be warmed by the hands. If the mercury fails to rise within 6 hours the reason is probably to be found in the cork fitting badly. For this reason it is perhaps advisable to run melted *wax-mixture*[2] round the line of contact between the cork and glass.

(3) *Sachs' method*[3].

Place 100 germinating peas in a jar, *A*, fig. 2, closed by an india-rubber cork pierced by two holes and fitted

---

[1] Or the beaker *Hg* may be filled with water.

[2] Wax-mixture consists of resin 15 parts, bees-wax 35 parts, vaseline 50 parts. The wax and the vaseline are melted together, the resin is powdered, gradually added and stirred.

[3] *Physiologie* (French Translation), 1868, p. 295, fig. 35. Also Pfeffer's *Physiologie*, I. p. 349, fig. 38.

with glass tubes. One tube is connected with an aspirator so that a current of air is drawn through the vessel and keeps up continuous normal respiration. The other tube serves to admit to the flask air free from CO_2; for this purpose it is connected with a filtering bottle F containing a few sticks of KHO[1]. The air is admitted to F through a tube T filled with soda-lime. To make sure that no extraneous CO_2 enters the flask, another washing bottle B containing baryta-water is fitted between F and the experimental flask, A. The drop-aspirator figured by

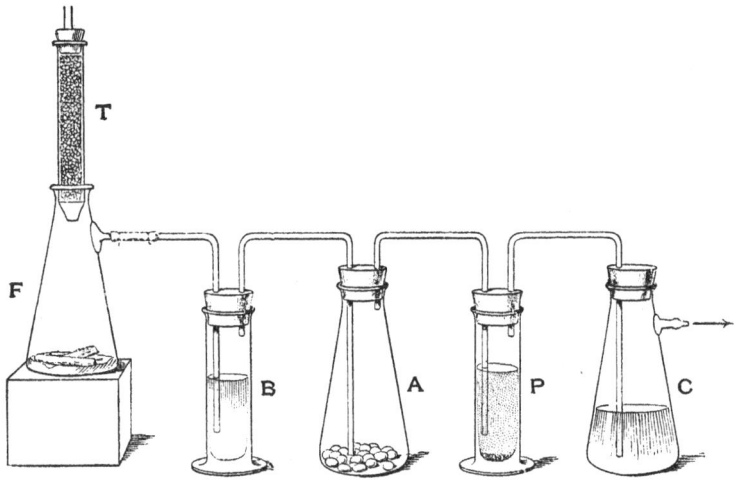

FIG. 2. Exp. 3.

Detmer[2] answers well; it is made from a distillation tube and is attached to a tap through which a current of water

[1] We now use a washing bottle containing KHO instead of the arrangement shown in the figure [1895].

[2] *Praktikum*, p. 179, fig. 76.

in detached drops passes, and produces a correspondingly slow suction-current of air at the side tube (c in Detmer's figure). The aspirator should be hung about 70 cm. above the table so as to allow the use of an outflow tube of about 60 cm. in length, which insures sufficient suction. The tube which admits air into F should be fitted with a screw clamp so as to regulate the inflow. If the sink in the laboratory is inconveniently placed the air-suction may be carried to any part of the room by means of fine lead-tubing.

Between the flask and the aspirator two washing bottles P, C, containing baryta-water are fitted in which the CO_2 produced by respiration of the plants is caught as $BaCO_3$. With 100 peas the amount of $BaCO_3$ may be estimated at intervals of 20 minutes or half-an-hour.

The estimation is made by titration, for which see Sutton, *Volumetric Analysis*, 5th Ed. pp. 80—89.

Rough quantitative determinations may be readily made as follows. Shake up about 21 grams crystallized Barium hydrate with 1 liter of distilled water and allow it to stand for 12 hrs. in a closed flask, or till dissolved. Filter it into a stoppered bottle. Before an experiment introduce, with a pipette, 50 c.c. of this solution into the bottle P, and a further quantity into C. At the end of the experiment, when a considerable quantity of Barium carbonate has been precipitated in P, draw off 20 c.c. of the solution [1] with a pipette and titrate it quickly against

[1] The liquid need not be clear, as extremely dilute HCl has no action on barium carbonate. For an accurate method of estimating the CO_2 of respiration by titration, see Blackman in the *Phil. Trans.* 1895, also a full account of his work in *Science-Progress*, 1895.

standard decinormal hydrochloric acid, using phenol-
phthalein (colourless with acids, pink with alkalis) as
an indicator.   Compare 20 c.c. of the original baryta
solution with the same acid.   The difference between the
amounts of HCl required to neutralize the two samples
gives a measure of the amount of baryta that has been
removed from each 20 c.c. of the liquid by combining
with carbon dioxide.   The whole amount of CO_2 produced
in the experiment is of course $2\frac{1}{2}$ times as great, and it
may be readily estimated from the following data:

$$1 \text{ c.c. } \frac{N}{10} \text{ HCl} = 0\cdot0022 \text{ gr. } CO_2 = 1\cdot19 \text{ c.c. } CO_2 \text{ at } 15° \text{ C.}$$

(4)   *Winkler-Hempel apparatus.*

By this apparatus the volume of CO_2 given off by
respiration in a known time may be fairly well determined.

To get a good result it is well to use a considerable
quantity of material, say 200 peas just beginning to
germinate.   They are placed in a conical flask of 400 c.c.
or 500 c.c. capacity and closed by a good rubber cork.
After 1 or 2 hours a sample of gas is drawn off for analysis.
This may be managed by an arrangement similar to that
shown in fig. 3, in which water flows from *o* to *i* as the
gas is withdrawn.   The arrangement figured was meant
for experiments on assimilation when several determinations of the CO_2 in the jar *J* are made: in the present
experiment the test-tube *i* is omitted and the water flows
into the bottom of the flask.   Thus only a single sample
of gas can be analysed with accuracy since the water

introduced into the flask absorbs the CO_2 produced. A second sample of the gas may however be used if it is taken within a few minutes of the first. When the test-tube i is omitted the flask can be shaken before analysis so as to insure that there is no accumulation of CO_2 at the bottom.

The analysis is made in the following manner:—A strong KHO solution (1 in 2) is introduced into B (fig. 4)

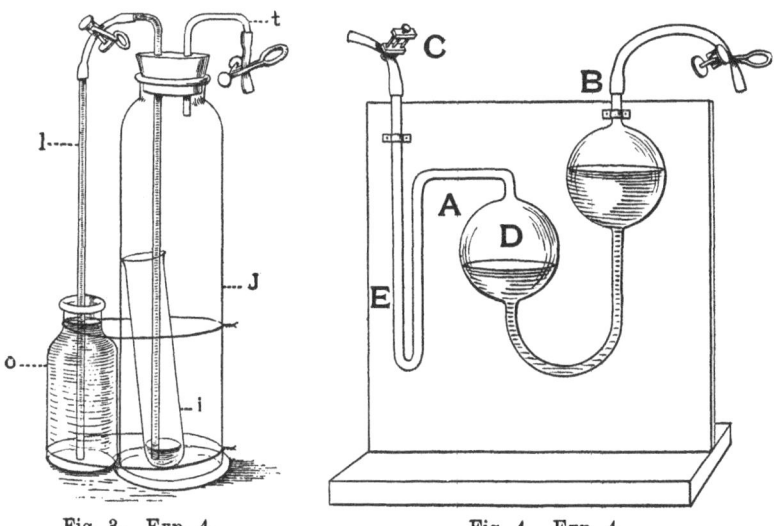

Fig. 3. Exp. 4.   Fig. 4. Exp. 4.

until its level reaches A, and then by blowing down B the KHO is forced up the fine tube E and into a thick-walled india-rubber tube connected with it. As soon as the solution appears at the open end of the tube, the clamp C is closed. The tubes G and F (fig. 5) of the measuring burette are then a little over half filled with

distilled water, care being taken that no air bubbles
remain in the connecting india-rubber tube. *F* is then
raised till water flows out of *H*; then the stop-cock *L* is
closed and *H* is connected by tubing with the vessel *J* in
fig. 3 containing the gas to be analysed. *F*, now nearly
empty, is lowered and *L* opened, so that a sample of gas
is drawn into the burette. *L* is closed and *H* disconnect-
ed. The volume drawn in is then measured by means of

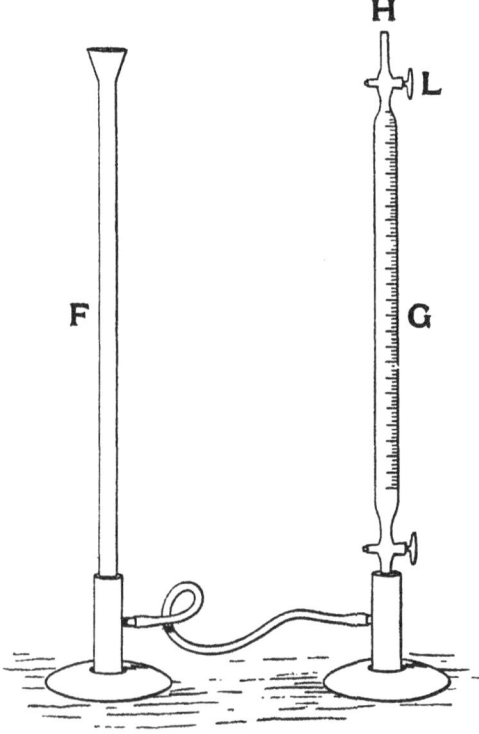

Fig. 5.   Exp. 4.

the graduations on G, after bringing the water in the two
tubes to one level. To absorb the CO_2, H is connected
with the india-rubber tubing C of the absorption pipette
(fig. 4). F is raised, and L and the clamp C opened. The
gas is thus forced over into D, where it is retained for a
minute or so and gently shaken in contact with the KHO,
the clamp C and stop-cock L being closed meanwhile.
When absorption is believed to be complete the gas is
sucked back into G (fig. 5) by lowering F, C and L being
open. L is then closed and G and F brought to a level
so that the diminished volume of gas can be again read
off. The difference gives the amount of CO_2 originally
present. To make sure of complete absorption the gas
may be again passed into D, shaken and returned, when
it should show no further reduction in volume.

When any potash is sucked back into G along with
the gas the tubes must be carefully washed clean before
being used for another sample of gas.

(5) *Timiriazeff's Eudiometer.*

The modification of Timiriazeff's eudiometer which
we employ for the analysis of gas given off by assimilating
water plants (see exp. 52, p. 45) may be conveniently
employed for respiration experiments. Five or six germinating peas are placed in a test-tube (15 c.c. capacity)
which is tightly closed by means of a rubber cork. It
may be inverted in water or mercury, and the contained
gas analysed after two hours. For this purpose it is un-
corked while the mouth of the test-tube is under water and
about 0·5 c.c. poured up into the funnel of the eudiometer.

In one of our experiments we found the end of the inflorescence of *Reseda luteola* convenient material, yielding a good percentage of CO_2.

(6) *Intramolecular respiration.*

To demonstrate the fact, the following simple form of experiment may be tried.

Soak 6 peas in water for 12 hours, when the seed-coats can easily be removed without injury to the embryo; the removal of the testa is necessary to avoid introducing air with the peas, the object of the experiment being to show that CO_2 is produced in the absence of free oxygen. Fill a test-tube with mercury and invert it in a mercury trough, which should stand in a strong wooden tray. This precaution is advisable in all experiments involving the use of mercury, so that if any accident occurs the mercury may not escape and get into the cracks of the floor.

Pass the peeled peas one at a time under the rim of the test-tube so that they float up into the mercury, and occupy the upper end of the test-tube. On the following day it will be found that the test-tube is half full of gas, and the peas are therefore clearly visible, instead of being partly hidden by mercury.

A few drops of water are now passed in under the test-tube rim with a bent pipette, and a fragment of caustic potash added from below, in this way a strong solution of KHO is supplied, by which the CO_2 is absorbed.

(7) *Another method.*

The Torricellian vacuum was used by Wortmann in

his work on intramolecular respiration[1]. A tube closed at one end, and of greater length than the height of the barometric column, is filled with, and inverted over mercury. Three or four peas are floated into the vacuum at the top of the tube. After 24 hours a depression of several cm. will be observed in the height of the column, which will, on the addition of KHO, rise to about its original level, allowance being made for any change in the barometer.

(8) *Pfeffer's method.*

To get accurate results another method must be followed; the following is taken from Pfeffer's paper in his Tübingen *Untersuchungen*, Vol. I. p. 637.

The principle is that described in exp. 3 for estimating ordinary respiration, but instead of air, a current of hydrogen is drawn through the vessel in which the plants are contained. It is necessary to prevent the entrance of extraneous CO_2 and to make sure that the hydrogen has no admixture of oxygen.

(9) *Rise of temperature*[2].

The following is Sachs' arrangement for showing the rise of temperature in germinating peas[3]. We have found flowers, such as those of the dandelion (*Taraxacum*),

[1] Sachs' *Arbeiten*, II. p. 500.

[2] The spadix of *Arum* is the classical material for demonstrating the heat produced by the respiration of plants. The rise of temperature only occurs during such a limited time that the experiment constantly fails and is not to be recommended for class-work.

[3] Sachs' *Text-book*, Ed. II. p. 724. Fig. 472.

treated in the same way to answer well. Gather a large handful of dandelion flowers[1], cutting the stalks just below the head, place them in a large funnel supported in a beaker half filled with KHO. Hang a thermometer so that the bulb is covered by the flowers, and let the control thermometer be supported in a funnel containing coarse sawdust slightly moistened and loosely packed. This arrangement is meant to equalise the conditions of the two thermometers, and to prevent the thermometer among the flowers acting as a wet-bulb. We find, with the control thermometer hanging simply in the air, that the flowers keep about 2° C. above the control temperature. As before, the whole must be covered with a bell-jar. Sachs uses a tubulated bell of which the opening is plugged with cotton-wool[2].

(10)   *Oxygen necessary.*

Several of the earlier observers have shown that when the air is replaced by indifferent gas the temperature falls. Pfeffer[3] recommends that the germinating seeds or other material should be placed in a glass balloon having three apertures—one of which serves for a thermometer. When the temperature of the respiring material

---

[1] Or in winter of young flowers and buds of a small-flowered *Chrysanthemum.*

[2] To demonstrate the heat of respiration, Professor Errera of Brussels (as he is good enough to inform us) uses a Leslie's Differential Thermometer. One of the air-bulbs is plunged in a funnel filled with germinating barley and covered loosely with damp paper. The other bulb is in a similar funnel containing killed barley.

[3] See Pfeffer, *Physiologie*, II. p. 403. Fig. 40.

has been proved to be steadily above that of the surrounding air, the atmosphere in the balloon is replaced by hydrogen, for which purpose the two lateral apertures will serve. The readings of the two thermometers should now become practically equal, and according to Pfeffer it is possible to re-establish the difference by readmitting air. The experiment is a difficult one and should only be attempted by a student of some experience.

(11) *Succulents* [1].

In certain succulents an increase of the acidity of the cell sap is accompanied by a fixation of oxygen.

A leaf of *Rochea falcata* is taken from the plant at the close of a hot summer-day, cut into pieces and introduced into a graduated gas-tube of simple test-tube form: it may be kept in place by a plug of glass-wool. It is not necessary to stand the tube in mercury, water will serve quite well. We have also employed an arrangement like that given in fig. 1, the KHO being omitted and water replacing the mercury in the beaker *Hg*. The apparatus is kept in the dark until the following morning, when a considerable rise in the water column is visible. As a control a similar graduated tube is fitted up with non-succulents, such as pieces of young sunflower [2].

[1] See De Saussure, *Recherches Chimiques*, (An. xii = 1804), p. 65 ; also Detmer, *Praktikum*, p. 224, whose arrangement of the experiment we have adopted.

[2] To complete the experiment, the relative acidity of the *Rochea* in the evening and next morning should be compared. See Part II.

SECTION B.   **The effect of various temperatures :
of certain poisons : and of electrical shock.**

(12)   *Temperature* [1].

To get a rough idea of the upper limit of temperature
which ordinary plants can endure, it is well to make a few
simple experiments with plants in which the moment of
death is marked by some obvious change, e.g. in colour.
Oxalis acetosella is useful for this purpose, because death
is indicated by a dingy yellow colour due to the action
of the acid cell sap on the chlorophyll.

Fill a beaker with water at 25° C., and suspend in it
a thermometer, to the bulb of which a leaf of *Oxalis* is
attached.   Heat the water by means of a gas flame, and
note the temperature at which the leaf loses its fresh
green tint.   The colour begins to change at about 52° C.

(13)   *Temperature.*

If the *Oxalis* leaf is previously injected with water
under the air-pump, it changes colour at a temperature
several degrees lower than in exp. 12.   This is a simple way
of demonstrating the fact given by Sachs (*Physiologie*,
p. 71) that plants in air endure a temperature which they
cannot bear in water.

The cells of the injected *Oxalis* leaf acquire the temperature of the water more quickly than those of the
uninjected leaf, and this is probably the explanation of the
difference.

[1] See Chapter III., on the general conditions of plant life, in Sachs'
Text-Book of Botany, Edition II., also his *Physiologie* (French Translation), p. 56.

(14) *Temperature.*

When a turgid cell is killed, the cell sap escapes through the dead protoplasmic wall, and if the cell sap is coloured, the escape will be a marked occurrence. The Beet-root (*Beta*) may be used in this way as a rough indicator of the temperature at which the protoplasm is killed. Cut a slice of beet-root, 3 or 4 mm. in thickness, wash it to free it from any cell sap adhering to the cut surfaces, and suspend it with a thermometer in a beaker of water at about 25° C., which is to be heated as in experiment 12, but the temperature should be allowed to rise very slowly. A temperature of 55° or even 57° will be required.

A similar experiment may be more accurately made under the microscope, using one of the methods described below, by which a microscopic object can be subjected to a given temperature.

(15) *Dry and soaked seeds*[1].

The effect of a high temperature depends, among other things, on the condition of the subject of the experiment. Thus, dry seeds can endure a temperature which is fatal to seeds which have been soaked.

Take 20 peas, half of which (*a*) are to be left in water for 12 hours, or until they are thoroughly soaked, while the other 10 (*b*) are reserved for comparison. The dry seeds (*b*) are placed in a dry test-tube, while the imbibed seeds (*a*) are placed in a test-tube half full of water: both

---

[1] Sachs' *Physiologie* (French Tr.), p. 72.   Fig. 8.

test-tubes are corked and are immersed in a beaker of
water kept by means of thermostat at 60° C. for 2 hours.
Both sets of seeds should now be sown in damp sawdust,
—the lot (*b*) having been previously soaked in cold water
for twelve hours : it will be found that lot (*b*) germinate,
while (*a*) do not do so, and show other obvious signs of
being dead.

(**16**) *Circulation of Protoplasm—Sachs' Hot-box.*

Any parts of plants, in which circulating protoplasm
can be observed, serve as material for studying the effects
of temperature. The staminal hairs of *Tradescantia* or
other plant-hairs are convenient, or the tentacles of *Drosera*
may be used. But the leaves of *Elodea* are perhaps most
easily obtainable throughout the year.

Mount a leaf of *Elodea* upside down in a drop of water
under a large cover-glass ; look for circulating protoplasm
near the mid-rib[1], and subject it to gradually increasing
temperature by means of any of the recognised " hot-
stages," e.g. with Sachs' Hot-box. The arrangement is
described and figured in Sachs' *Text-Book*, English Translation, p. 736. It consists of a hollow-walled metal box,
into which the microscope is placed so that by filling the
walls with warm water the object under observation can
be subjected to the desired temperature. A window
admits light, and a hole in the moveable lid allows the
microscope-tube and fine adjustment to project. The

---

[1] The leaves should be cut off an hour before they are wanted,
because, in winter at any rate, circulation is not visible until some time
after the leaves have been cut.

felt lining of the lid should be wetted and a little water should be spilt on the floor of the box, so that the atmosphere surrounding the object may be damp. A thermometer passes through a hole in the lid or, as we find more convenient, through a cork fitting one of the lateral openings. The glass slip on which the object is mounted should be separated from the stage of the microscope by a perforated plate of cork, so that the object may assume the temperature of the air, rather than that of the microscope,—although these two temperatures will after a time be nearly identical.

The hot-box may be conveniently supported on wooden blocks and heated by a gas flame. As the warming of a considerable mass of water is a slow process it is advisable to fill the box with water 10° C. above the room temperature. Notice the accelerating effect of warmth, and record the temperature at which the circulation (1) becomes slower (42°—46° C.); (2) stops altogether (about 46°—50° C.).

(17) *Velten's method* [1].

A simpler and quicker plan is that of Velten, which, however, should not be used with a valuable microscope. The objective and the preparation are immersed in water contained in a glass dish standing on the stage of the microscope. A siphon, provided with a tap, allows warm water to run into the dish, while a second siphon and tap

---

[1] *Flora*, 1876, p. 177, also F. Darwin, *Q. Journal Microscopical Science*, N.S. Vol. XVII. p. 245. Cf. Pfeffer, *Zeitschr. für wiss. Mikroskopie*, 1890.

provides for the overflow. The object is mounted between two cover-glasses, which are gently clamped together by a

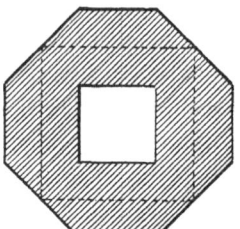

FIG. 6. Exp. 17.

bit of tinfoil of the form shown in fig. 6, the flaps being bent up at 45° along the dotted lines. Unless some such plan is adopted, the upper cover-glass is liable to be washed off by currents in the water.

The same method may be used to subject circulating protoplasm to a low temperature.

(18) *Effect of CO$_2$.*

To observe the effect of gases on circulating proto-plasm, the *Elodea* leaf is mounted in a small drop of water, on the under surface of a cover-glass forming the roof of a gas chamber: if the cover-glass projects fairly well beyond the edges of the hole on which it lies, the apparatus can be made sufficiently gas-tight by painting the edges of the cover-glass with olive oil; or the glass may be fixed with putty. Having under observation a circulating cell, attach the tube of the gas chamber to the CO$_2$-generating apparatus[1], and observe that the proto-

---

[1] The CO$_2$ must be made to bubble through water before it reaches the gas chamber.

plasm comes to rest: by disconnecting and allowing air to pass, the circulation can be renewed.

In this experiment the CO_2 acts, not by preventing access of oxygen, but as a narcotic. This can be shown by connecting with a hydrogen-generator; the rapid retardation previously observed will be absent.

(19) *Chloroform.*

The same apparatus serves to demonstrate the effect of chloroform and other hurtful vapours. Shake up one per cent. of chloroform in a bottle of water, through which (by means of an aspirator) a current of air is made to bubble. The air, thus charged with chloroform is allowed to pass through the gas-chamber. The circulation may be stopped without killing the leaf.

(20) *Chloroform.*

The effects of poisons may also be conveniently demonstrated on the leaf of *Oxalis acetosella,* using the colour test already described.

Shake up 1 c.c. chloroform in 200 c.c. water in a stoppered bottle and add an *Oxalis* leaf cut into small pieces. Note the time required for the discoloration to occur.

(21) *Carbolic acid. (Phenol.)*

Make the same experiment, substituting 0·5 per cent. carbolic acid for chloroform-water.

(22) *Tetanising current.*

If an *Oxalis* leaf is impaled on a pair of needles (in an insulated handle) connected with the induction coil, the region between the punctures is killed and becomes discoloured when the current passes : the needle points should not be more than 2—3 mm. apart.

(**23**) *Tetanising current.*

Two triangles of platinum foil are sealing-waxed on to a glass-slip, the points being about 1 mm. apart. To make the platinum adhere well it is necessary to heat the glass over a flame until the wax between the glass and the metal is thoroughly soft, and then to apply pressure. An *Elodea* leaf is mounted in water so that a cell, showing circulation, lies between the points, and by connecting the foil triangles with an induction coil, the effect of the tetanising current can be observed. The wires from the coil are most conveniently connected by means of the insulated screw-binders, obtainable from instrument makers ; in the absence of screw-binders the following arrangement will be found to answer quite well. A cork ring is sealing-waxed on to each foil-triangle near its base, and into the little vessels so made, mercury is poured, into which the connecting wires are placed. To get a rough idea of the current needed, it is advisable to note the position of the coil when the current is just bearable on the tongue, and compare it with the position of the coil when the protoplasmic circulation has been stopped.

CHAPTER II.

SECTION A. *Formation of Starch.* SECTION B. *Evolution of Oxygen.* SECTION C. *Reactions of Chlorophyll.* SECTION D. *Conditions of Chlorophyll formation: Etiolation: sun and shade leaves.*

SECTION A. **Formation of Starch.**

(24) *Sachs' Iodine-method*[1] (*Iod-Probe*).

This is a macroscopic method well adapted for many experiments. Almost any leaves will serve as material for the demonstration of the method, but since in research it is of importance to employ material which allows of rapid work, the choice of plants is a point to be considered. Submerged water-plants are useful, and among land plants, *Tropæolum* and clover (*Trifolium*) are especially valuable. The leaves to be tested are to be boiled for about one minute in water[2], when they should be flaccid

[1] Sachs' *Arbeiten*, III. p. 1.

[2] Sachs allows a longer period, viz. 10 minutes; he states also that the addition of a few drops of strong KHO to the boiling water hastens the process.

and free ·from intercellular air. They are then placed
in alcohol warmed to 50°—60° C.: cold alcohol will
remove the chlorophyll equally well but not so quickly : if
the specimens are not wanted at once the best results
will be obtained by putting them in the sun for a few
hours. The preliminary boiling in water must on no
account be omitted, it shortens the process of decolorising
in the most remarkable manner; of this it is easy to
convince oneself by trying, for instance, to decolorise an
Enteromorpha without the hot-water treatment. To
produce the iodine reaction, place the decolorised leaves
in alcoholic tincture of iodine diluted with water[1] to the
colour of dark beer. In a few minutes they will be
stained, and after washing in fresh water, they should be
spread out on a white plate so that their tint—by which
the amount of starch is roughly gauged—may be well
seen. When full of starch they are almost black, and
with less amounts of starch the colour sinks through
purple, grey, and greenish grey to the yellow tint of
starchless leaves.

(25)   *Schimper's method*[2].

In some cases it is necessary to use the microscope,
this is especially necessary when the amount of starch
present is small, or where, as in Schimper's researches,
the distribution of starch in the leaf is minutely
studied.

Prepare a strong solution of chloral hydrate by dis-

[1] It is not necessary to use distilled water.
[2] *Bot. Zeitung*, 1885.

solving the crystals in as much distilled water as will just
cover them[1]. The solution is now coloured by the addition of a little tincture of iodine, and is ready for use.
Delicate leaves, such as those of submerged water-plants,
when placed in Schimper's solution, are rendered so transparent that every detail of starch-distribution can be
studied in the leaf examined as a transparent object under
the microscope.

(**26**) *Variegated leaves.*

Test Sachs' method on a variegated leaf such as that of
the ivy (*Hedera*) or of *Arundo donax*. In the case of the
ivy a rough plan of the green and white parts of the leaf
must be traced on paper placed under the leaf, which may
best be done by tracing a broken line with a blunt instrument dotted along the lines separating the chlorotic from
the green parts of the leaf. The iodine-stained leaf is
then compared with the plan. With *Arundo* no such
process is necessary, the chlorotic regions are in longitudinal stripes, and it is only necessary to cut out of the
leaf a short piece, which, after staining in iodine, can be
replaced between the base and apex of the leaf to which
it belonged : the colourless stripes in the fresh parts cor-
respond to yellow stripes in the stained part, and the
purple to the green. Both the extraction of the chlorophyll and the staining with iodine are slow processes in
the case of *Arundo*.

(**27**) *Disappearance of starch in darkness.*

Either of the methods may be tried on submerged

[1] Chloral hydrate 8 parts, water 5 parts.

water-plants (e.g. *Elodea, Potamogeton*) which have been placed in the dark room for about four days. The control-plants must be grown either out of doors or in a greenhouse.

(**28**) *Effect of dull light.*

Sachs' method may be used to demonstrate a fact, the knowledge of which is of practical value to the physiologist[1], namely, that plants in a laboratory suffer from want of light far more than would be readily supposed—and that accordingly experimental plants cannot be too carefully kept in the best light available.

Choose two equally vigorous pots of clover, let one remain in bright diffused light out of doors, and place the other on a table in the middle of the laboratory. The plant in the laboratory must be under a bell-jar on account of the dryness of the air, and therefore to make the control experiment fair the plant out of doors should also be under a bell. After two days compare the amounts of starch in the two plants.

(**29**) *Local effect.*

Various means may be used to convince oneself that assimilation is confined to the illuminated regions of a leaf. Part of a leaf may be darkened, while still attached to the plant, by bending it down and burying the apical half in a flower-pot of finely sifted dry earth. The leaf should be buried one day and examined in the afternoon of the following day, taking care before the leaf is uncovered to mark on it the depth to which it was buried.

[1] See Detlefsen. Sachs' *Arbeiten*, III. p. 88.

(**30**) *Gardiner's experiment* [1].

A plant growing in a flower-pot (for convenience of moving) is placed in the dark for 24 hours, or until the leaves are found to be free from starch. One of the leaves is now covered with a photographic negative and left exposed to bright light out of doors, or in a greenhouse, until the evening, when the leaf is tested for starch. It will be found that an accurate copy of the photograph has been printed in starch.

(**31**) *Effect of rays of different refrangibility.*

The effect of the different parts of the spectrum may be demonstrated by a method similar to that described in Exp. 29, as has been done by Timiriazeff [2]. In the absence of the necessary apparatus we may compare the effects of light transmitted through coloured fluids. Fill a couple of double-walled bell-jars, (1) with potassium bichromate solution, (2) with ammoniacal $CuSO_4$ solution. Under each bell place a young *Tropæolum* or clover plant in a small pot, or a seedling plant of any kind dug up and placed with its roots in a bottle of water. The bell-jars should stand in saucers of dry earth or sawdust, so as to ensure the exclusion of colourless light. They must be exposed to diffused light—in sunshine the temperatures are not the same in the two bell-jars. The exposure should be for $1\frac{1}{2}$ or 2 days. The plants in the blue light will be almost starchless.

---

[1] W. Gardiner, *Annals of Botany*, IV. p. 163.

[2] Timiriazeff, *Comptes rendus*, T. CX. p. 1346.

(32)   *Terrestrial leaves under water.*

To show that the leaves of land-plants do not form
starch as those of aquatic plants do under water[1], it is
only necessary to tie a leaf so that it is partly immersed
in a beaker of water. The experiment may be started
in the morning and concluded on the afternoon of the
following day.

(33)   *The importance of the stomata in supplying the
         path for gaseous exchange*[2].

For this experiment leaves should be employed in
which the stomata are all on the lower surface. Stahl
uses *Prunus padus*; we find *Sparmannia africana* gives
good results, and no doubt many other plants would
answer the purpose. The lower surface of one half of a
leaf is carefully painted with vaseline, or as Stahl recommends with melted cocoa-fat and beeswax. The
plant having been exposed to a good light for two days,
the leaf is subjected to the iodine test. The painted
half (in which the stomata are blocked) will be either
quite or nearly starchless, while the control half shows a
normal amount of starch.

(34)   *Effect of excess of CO_2.*

To show that excess of CO_2 diminishes assimilation[3]
floating water-plants are convenient. We use *Callitriche*,
and possibly *Lemna* might be used, but these must be

[1] Nagamatz (Sachs' *Arbeiten*, III.) shows that leaves covered with
bloom can assimilate under water.

[2] Stahl, *Botan. Zeitung*, 1894. See also F. F. Blackman, *Phil. Trans.*
1895, and in *Science Progress*, 1895.

[3] Godlewski. Sachs' *Arbeiten*, I. p. 343.

kept a long time in the dark before they are destarched. Two graduated jars of 200 c.c. capacity are filled with and inverted over water, and plants of *Callitriche*, which have been previously deprived of starch, are passed under the edge and allowed to float up. Into one jar equal quantities of air and CO_2, while into the other 12 volumes of air to one of CO_2 are passed. The proportion of CO_2 in the atmospheres so prepared does not of course remain constant, since the water absorbs the gas. But if the experiment is started in the evening and concluded in the evening of the next day, one jar will certainly contain far more than the optimum of CO_2, while the other will not fall much below the optimum. A still simpler plan is to use beakers of about 800 c.c. capacity inverted in saucers of water. The beakers are graduated as follows: into one 550 c.c. of water is poured and the level marked with a diamond, a second mark being made after the addition of 50 c.c. The other beaker is marked at 300 and 600 c.c. The beakers are filled with water and inverted in saucers, and the rosettes of *Callitriche* floated up under the rims of the beaker. Three hundred c.c. of air are now introduced into one beaker and 550 c.c. into the other, using a finger bellows for the purpose; afterwards CO_2 is added until each beaker contains 600 c.c. of mixed gas, one containing 50 p.c., the other 8 p.c. of CO_2. In our experiments the *Callitriche* exposed to 50 c.c. CO_2 showed hardly any starch, while the control-plants were black with it.

The experiment may be more accurately performed with a pair of graduated tubes inverted over mercury (covered with a few drops of water) and containing leaves of land-plants.

(**35**) *Plants deprived of CO_2.*

To show that the formation of starch depends on the presence of CO_2 it is necessary to cultivate plants in such a way that they have access to oxygen but not to CO_2.[1]

Water-plants.

Water which has been boiled and allowed to cool in a

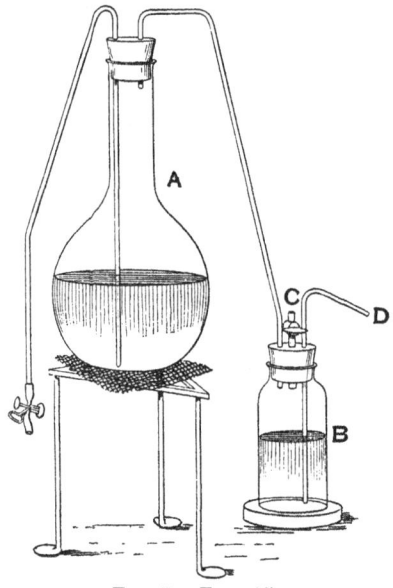

FIG. 7.   Exp. 35.

closed flask will be free from both O and CO_2. But if the flask is connected with an arrangement preventing

[1] Godlewski, *Flora*, 1873, p. 378.

the access of CO_2 while allowing other gases to pass in, the boiled water will after a time become oxygenated.

A convenient method is the following. A flask A (fig. 7) is filled with spring water which has been freshly boiled, and filtered from precipitated calcium carbonate; it is connected with the bottle B, half filled with strong KHO solution. The water in A is boiled 20 minutes, with the stop-cock C left open. The flame is now removed and C is closed. As the flask A cools, air is sucked in by D, and in passing through the KHO in the bottle B, is freed from CO_2. The water so prepared is now used for the culture fluid: the vessel containing the plants must be closed by a rubber cork through which passes a tube of soda-lime like the one shown in fig. 8.

A similar flask filled with spring water (to which a little extra CO_2 may be added by blowing air from the lungs through it) and closed by a U tube containing coarse sand, will serve for a control.

The CO_2 may also according to Pfeffer [1] be removed by careful treatment with lime water.

Land-plants.

Seedlings with their roots in water, or plants of *Tropæolum* or clover in small pots, are to be used. The pot is supported in a crystallising glass (G, fig. 8) half filled with soda-lime, which rests on a ground glass plate, and is covered by a tubulated bell-jar, the lower edge of which is ground, but need not be welted. The ground edge is

[1] Pfeffer, *Physiologie*, I. p. 111.

smeared with wax-mixture, and the junction with the glass plate is made secure by a little embankment of

FIG. 8.   Exp. 35.

wax-mixture melted into the angle with a hot wire. The aperture of the bell is closed by a rubber cork pierced for the tube T, which contains soda-lime.

The apparatus should be placed out of doors or in a brightly lighted greenhouse. A control-plant must be fitted up in a similar way except that G may be dispensed with and that T must be filled with sawdust or some indifferent coarsely grained powder. We find that exposure from 10 a.m. until the afternoon of the next day gives good results.

(**36**) *Gain in weight.*

Sachs[1] has shown that a given area of leaf is heavier in the evening than in the morning, owing to the accumulated products of assimilation.

The following are Sachs' instructions for performing the experiment. Out of a board 3 mm. in thickness cut out a square of 10 cm. to the side and another rectangular piece of 10 × 5 cm. : these are to be used as templates by which to cut out areas of 100 sq. cm. and 50 sq. cm. respectively. The plants used must be large leaved kinds, e.g. *Helianthus, Cucurbita, Rheum.* The experiment must be begun soon after sunrise[2]. Five or six healthy leaves having been selected, each must be divided longitudinally close to one side of the midrib; the part which is thus freed from the plant is to be investigated at once, while the other half remains on the plant till the evening. Each half-leaf is treated in the following way. It is laid on a flat board, the lower side of the leaf being upwards, so that the projecting veins may be easily seen. The templates are now fitted in between the larger veins so as to get areas as free as possible from large veins. The rectangular pieces of leaf so obtained are quickly killed by steam. After being allowed to become air dry, they are powdered, dried, and weighed.

In the evening a similar process is gone through with the control halves. The following is the result of one of Sachs' experiments. A hundred sq. cm. were cut out of

[1] *Arbeiten*, III. p. 19.
[2] Unless the plant is placed in a dark room on the previous evening, in which case the operator chooses his own time in the morning.

the halves of 7 leaves of *Helianthus annuus*; the dry
weight of the 700 sq. cm. was:—

5 a.m.   3·054 grams.

3 p.m.   3·693   ,,
          ————
          ·639   ,,

This equals 0·9 grams per sq. meter of leaf surface, per hour.

Mutatis mutandis the weighing method is used by
Sachs for showing the loss by translocation in the night.

(**37**)   *Translocation.*

Sachs' iodine method is also useful for studying the
translocation of carbohydrates, i.e. that the products of
assimilation wander from the leaf to the body of the
plant[1].

In the evening remove the halves of several leaves and
having tested small pieces of each (which should be
preserved for further comparison) place the freed halves
on wet filter-paper under a bell-jar in a cool dark room;
the plant must also be placed under a bell in the same room.

In the morning the half-leaves attached to the plant
will have lost more starch than the free halves. We have
found *Sparmannia* give a good result when darkened from
5 p.m. until 10.30 a.m., the half-leaves attached to the
plant being starchless and contrasting well with the free
halves.

(**38**)   *Assimilation of sugar*[2].

Water-plants, such as *Elodea, Potamogeton, Lemna,*

---

[1] More accurate methods are described in Part II. Chaps. xiii. and xiv.

[2] Böhm, *Botan. Zeitung*, 1883 ; Meyer, *Botan. Zeitung*, 1886 ; Acton,
Proc. Royal Soc. Vol. 47.

or *Callitriche*, are placed in vessels of 500 c.c. capacity, containing spring water, to one set of which (A), 3 % cane-sugar has been added, to (B), 5 % glycerine, while to (C) nothing has been added. It is of importance that specimens similar in size and in general vigour shall be selected, and that the specimens should be small in comparison with the volume of water in the beaker. Leave the vessels in the dark room for 8 or 10 days [1], when the plants in (A), (B) and (C) are to be compared as to condition, growth, and especially as to the contained starch. The control specimens will be starchless, and dead or nearly so, while the experimental plants will be obviously better nourished and will contain more or less starch. The glycerine cultures do not as a rule succeed so well as those in cane-sugar. The chief difficulty experienced is the growth of moulds in the solution. Something may be done by washing the vessels with ½ p.c. corrosive sublimate and then in boiled distilled water; the culture fluids should be boiled and allowed to cool in vessels closed with cotton-wool plugs. [See Chap. iii.]

Chlorophyll is not necessary for this form of assimilation, colourless parts of plants form starch vigorously. The white flowers of *Phlox paniculata* are especially useful for this experiment. They are simply floated in the above-described solutions of sugar or glycerine, control specimens being placed in water. In a few days they become rich in starch, while the control flowers are starchless. The employment of colourless objects, such as white flowers, is especially convenient, since the use of

[1] In summer Lemna shows an excellent effect in 6 days.

alcohol as a decoloriser is avoided. The flowers must, however, be boiled before being placed in the iodine fluid.

(39) *Formaldehyde.*

Loew[1] and Bokorny[2] have shown that although formaldehyde is poisonous even in very dilute solutions yet that oxymethyl sodium sulfonate (which is easily decomposed into formaldehyde and $NaHSO_3$ can be used in culture fluids, in the proportion of 0·1 per cent., without injury to *Spirogyra*. Bokorny (*loc. cit.*) has shown that if *Spirogyra* is cultivated, in the light, in a nutrient solution containing 0·1 per cent. oxymethyl sodium sulfonate, the starch in the plant increases considerably, a result which we have confirmed. The access of CO_2 must of course be prevented: for this reason the cultures must be examined for moulds, or bacteria which might serve as a source of CO_2 to the algæ. The nutrient solution must contain 0·1 per cent. disodic phosphate to counteract the evil effects of the $NaHSO_3$ set free. After four or five days the plants must be compared with the control specimens which have been grown under identical conditions but without oxymethyl sodium sulfonate.

(40) *Starch-formers (leucoplasts).*

These may be examined in the tubers of *Phajus grandifolius,* according to the method given by Strasburger[3]. The sections are to be placed in alcoholic tincture of iodine diluted with half its volume of distilled

---

[1] *Botan. Centralblatt,* XLIV. p. 315.

[2] *Berichte d. D. Bot. Ges.* IX. p. 103.

[3] *Praktikum,* pp. 67, 68.

water. The relative positions of starch-former and starch-grain and the elongated crystalloid are well shown in Strasburger's figure 29. The leucoplasts in the rhizome of *Iris germanica* are given in his fig. 30.

SECTION B. **The Evolution of Oxygen.**

(**41**) *Bubbles of gas given off.*

Place a branch or two of a submerged water-plant, such as *Hottonia, Potamogeton crispus,* or *Elodea,* in a beaker filled with spring water which has been in the laboratory for 12—24 hours, and has acquired the temperature of the room. The cut ends of the plants must be upwards; and must be below the surface, to effect which it may be necessary to tie the specimens to a glass rod (see Pfeffer, *Physiologie,* I. fig. 17, and Detmer, fig. 12). The beaker is to be placed in sunlight, and evolution of gas from the cut ends of the specimens to be observed. To obtain a convenient series of small bubbles Pfeffer recommends varnishing the cut end of the shoot and pricking a fine hole in the membrane so produced. Select a branch which seems to be yielding a satisfactory amount of gas, and record, with a stop-watch, the time which elapses while 10 or 20 bubbles are given off. The observation must be repeated until the rate of bubbling is fairly constant. It is important to know that the evolution of bubbles of gas may be produced by other causes than illumination. Thus a plant which is exposed to feeble illumination and is not giving off bubbles may be made to do so by being transferred to a beaker containing soda-water freshly drawn from a "syphon." Devaux[1] has shown that this depends on the internal atmosphere rapidly assuming the gas-pressure of the water, by the diffusion of CO_2 from outside into the intercellular spaces. For the same reason, apparently, any movement of the water, e.g. stirring it with a glass rod causes an increase in the escape of gas. It is on account of this fact that we avoid the use of freshly drawn spring water, which has, in a less degree, the effect of "syphon" water on the yield of gas, and vitiates any inquiry into the causes which increase or decrease the rate of bubbling.

(42) *Light of different intensities.*

Now move the beaker into the shade, or cover it with a sheet of white paper, and take a fresh series of readings, and finally replace it in sunshine and record the rate once more. White paper placed round the beaker (which remains open above) may be expected to reduce the rate by about one-half. In the absence of sunshine, an incandescent electric light of 5—10 candle power may be used. We find however that the *Incandescent Gas Company's* burner is the most convenient source of artificial light. It is necessary to interpose a glass trough filled with water between the light and the plant, to prevent undue heating from the gas. For the same reason the water in the trough must be constantly renewed by means of a pipe connected with the water supply, and an overflow. With either the incandescent gas or the electric light the

[1] *Ann. Sc. Nat.* 1889.

intensity of illumination may be easily varied by placing the light at various distances from the plant. The variations in the rate of escape of gas do not however seem to be proportional to the changes in the intensity of light, as Wolkoff[1] found to be the case when diffused light was used for illumination[2].

(43) *Dependence on CO_2.*

Transfer the plant to a beaker filled with water which has been boiled in the apparatus shown in fig. 7, when the rate of bubbling immediately falls off greatly.

After a time the water may be supplied with CO_2 by blowing vigorously into it through a glass tube, when the evolution of gas increases in amount. As a check on the result the beaker should finally be placed in the dark, to make sure that the increased rate of bubbling is not a physical effect like that produced by effervescent water.

(44) *Temperature.*

Provide two beakers of water, one at a temperature of 24°—26° C., the other at 4°—5° C. Place a specimen in the warmer of the two and when the readings are constant transfer it to the cold water. In some of our experiments on *Elodea* we found that the escape of gas was immediately and completely checked by a change from 26° C. to 7° C. During the experiment take note of any changes in the brightness of the sky; if this precaution is forgotten it is easy to be deceived by a

---

[1] *Pringsheim's Jahrb.* v.

[2] See also Pfeffer, *Physiologie*, I. p. 208.

passing cloud causing an alteration in the rate of assimilation. For this reason it is better to use artificial light.

(45) *Chloroform*[1].

Repeat experiment 41 and add a small quantity (10 p.c.) of chloroform-water, that is, of water in which not more than 1 per cent. of chloroform has been shaken. If the experiment is cautiously performed it should be possible to seriously diminish the rate of gas-discharge without killing the plant.

(46) *Coloured light*[2].

Proceed as in experiment 41, and when constant readings are obtained, cover the beaker with a double bell-jar containing ammoniacal copper-sulphate solution and note the result. Sunlight must be employed as the source of light. After an interval of 4 or 5 minutes, when the readings should be approaching constancy, replace the blue jar by another containing potassium bichromate solution, and take a series of readings. It will probably be necessary to alternate the blue and orange light several times before a trustworthy result is obtained, otherwise it is impossible to make sure of avoiding the effect of slight variations in the intensity of the light.

(47) *Collection of the gas.*

Place a quantity of any of the above-named water-plants in a glass jar of about 12—14 cm. diameter.

[1] Bonnier and Mangin, *Ann. Sc. Nat.* 1886.
[2] Sachs' *Botan. Zeitung*, 1864.

Press the plants down into the water with an inverted funnel, which should be a large one, and should fit easily inside the jar; its neck should be cut short, so that the opening may be easily submerged. The gas given off by the plants will be guided by the funnel and may be collected in an inverted test-tube filled with water and placed over the opening. If the neck of the funnel is covered with one or two centimeters of india-rubber tubing, and if a test-tube be selected which fits tightly over the tube, no other support for the test-tube is needed. The funnel may be kept in its place by 3 bent glass rods attached to its neck and hooked over the rim of the jar.

When the test-tube is nearly full, the gas may be shown to be oxygen by the glowing of a splinter of deal which has been lighted, and is blown out just before it is thrust into the gas. The test-tube should be of such a size that it can be easily covered with the thumb.

(48) *Engelmann's blood method*[1].

A flask containing defibrinated bullock's blood is attached to the water air-pump and exhausted of oxygen. The fluid appears to boil during the process, which is assisted by the application of a temperature of about 35° C. The blood may be preserved in a venous condition, and may at the same time be charged with the necessary CO_2 by being tightly corked with a supply of carbonic acid.

Engelmann recommends a filament of *Spirogyra* about

[1] *Pflüger's Archiv*, Vol. XLII.

a centimeter in length, but for an obvious result a single leaf of *Elodea* or part of a *Hottonia* leaf seems to us preferable. It is mounted in a large drop of blood which is spread into a thick layer by supporting the cover-glass on strips of paper. The preparation is now exposed to sunlight for 3 or 4 minutes, or to bright diffused light for a somewhat longer period. The oxygen given off from the leaf brings back red arterial tint to the blood. The effect is perhaps most easily seen under a low power of the microscope, the red zone round the leaf giving the impression of a source of light hidden behind the leaf, producing in fact somewhat of a sunset effect. But it is also clearly visible with the naked eye, especially if the preparation is held above a sheet of white paper; it is not so plain if it lies on the paper. The venous blood has a dull, almost grey purple appearance in contrast to the red halo which surrounds and follows the curves of the leaf. The effect is quite clear and unmistakeable. A control specimen should be placed in the dark so as to make sure that the effect is not due to diffusion from the gas in the intercellular spaces of the leaf. Or the specimen which has been illuminated may be darkened, for about 1 hour, or until the red colour disappears, when the light effect may once more be produced.

According to Engelmann the most delicate method of showing the evolution of the oxygen is by means of the spectroscope, the spectrum of the blood changing as the oxyhæmoglobin appears.

To perform Hoppe-Seyler's[1] version of the experiment

[1] *Zeitschr. f. physiolog. Chemie*, Bd. II. p. 425, 1879.

(on which as Engelmann states he founded his method) the blood must be considerably diluted, and again exhausted. A sprig of *Elodea* is placed in a corked test-tube (see note 3, p. 51) and exposed to bright light with a control test-tube. The blood which contains the plant shows an arterial tint while the control blood remains venous.

(49)   *Boussingault's phosphorus method*[1].

Fill a bell-jar over water with hydrogen and add a small proportion of CO_2, i.e. not more than 8 per cent. of the volume.   Introduce a stick of phosphorus and a leafy branch.   The oxygen in the intercellular spaces of the plant will attack the phosphorus, and the bell-jar will be filled with white fumes.   The bell-jar must therefore be placed in the dark for two or three hours, or until the white fumes are dissolved in the water, and the contents of the jar are clear and transparent.   The bell-jar is now exposed to the sun, when in a few minutes it becomes clouded with white fumes.   We find that, when replaced in the dark, a quarter of an hour is sufficient for the absorption of the fumes.

(50)   *Pfeffer's method*[2].

A leaf is exposed to light in a calibrated tube containing a known volume of CO_2: after a certain number of hours the amount of CO_2 decomposed is estimated by absorbing what remains with KHO.   The tube is almost 36 cm. in length, of which 26 cm. is a calibrated tube of

[1] See Deherain, *Chimie Agricole*, p. 82.
[2] Sachs' *Arbeiten*, I. p. 15.   See also Pfeffer's *Physiologie*, I. p. 188.

14—15 mm. in diameter, T (fig. 9); above this part the tube is blown into a balloon and ends above in a narrow tube B with flat ground edges. The whole tube contains about 120 c.c. The leaf to be experimented on is rolled into a cylinder and gently pushed up the tube with a wooden rod until it reaches the wide part of the tube, where it unfolds of itself. After the experiment is over, the leaf is to be removed by means of a piece of thin iron wire, W, attached to the stalk before the leaf was inserted. During the experiment the wire should be attached outside the tube by an elastic band E. The tube is fixed vertically in a glass beaker, H, having upright sides and containing mercury, and a drop or two (0·2—0·3 c.c.) of water is placed above the mercury column in the calibrated tube to protect the leaf from mercury fumes. By applying suction at B the mercury column is raised to a desired height. The suction is best applied through a washing bottle containing water, so that the breath of the operator may not come directly in communication with the air in the gas-tube. An india-rubber tube fitting over B serves to connect with the washing bottle, and

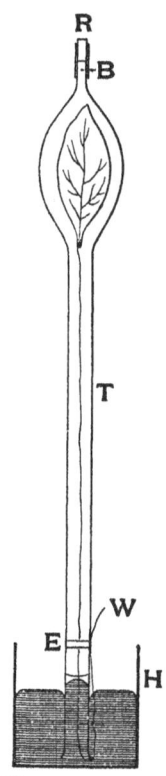

Fig. 9.   Exp. 50.
Copied from Pfeffer
loc. cit.

also to close the tube when desired.   When the mercury

column is at a sufficient height, the tube is temporarily closed with a clip and afterwards more securely by a bit of glass rod R, whose lower surface is ground flat and greased, so that when pushed home it fits close against the ground surface of B.

The height of the column of mercury is now read off on the calibrated tube, and at the same time the height of the little column of water is recorded. Readings of the barometer and thermometer are also taken. From 8 to 10 c.c. of CO_2 which has been washed in $NaHCO_3$ to free it from HCl is now passed into the tube, and the readings are again taken. Before introducing the CO_2 its purity should be tested by ascertaining that it is entirely absorbed over KHO. The apparatus is now exposed to bright diffused light for 5 or 6 hours, or it may be exposed to sunlight. When the exposure to light is complete the leaf must be pulled out by the wire, and when the apparatus has cooled, readings are again taken.

In order to estimate the quantity of CO_2 which has been decomposed, about 0·2 or 0·3 c.c. of concentrated KHO is injected into the gas-tube; this Pfeffer recommends to be done by the heat of the hand acting on a closed pipette. After 2 hours the CO_2 may be assumed to be all absorbed, when readings are again to be taken. The volume of the leaf is also to be ascertained by sinking it in a narrow measuring glass and reading off the altered position of the level; the fluid in the glass may be a mixture of alcohol and water which prevents adhesion of bubbles to the leaf. The volume of the leaf being known it must be applied as a correction to the readings of the gas-volume. To obtain the result it is necessary to reduce the readings of the calibrated tube (before and after the injection of KHO) to 0° C. and to 1 meter mercury pressure, and to make allowance for water vapour tension, etc. This is to be done according to the formula of Bunsen[1].

It is by no means necessary to employ a tube of the above-described form. We employ tubes of test-tube form of 2 cm. internal diameter and containing 100 c.c. Oleander leaves (which are especially good material in the winter months) fit these tubes well. The mercury is raised to the desired height by a thick-walled india-rubber tube pushed up into the cavity, and connected with a water air-pump. The rubber tube is then closed between the fingers and drawn out: if in this process a few drops of mercury are drawn into the tube, they may be sucked (by turning on the pump) into a bottle fitted like a washing bottle, which serves as a trap between the pump and any vessel to which suction is to be applied.

(51) *Winkler-Hempel apparatus.*

For demonstration purposes, where it is desirable to avoid barometer readings, calculations, &c., fair results

---

[1]
$$v_1 = \frac{(v - m)(b - b_1 - b_2)}{1 + 0 \cdot 00366 t^\circ}.$$

Where $v_1 =$ the reduced volume of gas.
$v \ =$ the observed volume.
$m \ =$ the correction for meniscus.
$b \ =$ the barometric reading.
$b_1 =$ the mercury pressure in the eudiometer.
$b_2 =$ the water-vapour tension at the temperature t°.
See Bunsen and Roscoe, *Gasometric Analysis.*

may be obtained with the Winkler-Hempel apparatus,— described in exp. 4, p. 6.

The jar, *J*, fig. 3, containing leaves is filled with air containing about 8 °/₀ of CO_2: the *exact* proportion is of no importance, but it must be accurately determined at the beginning of the experiment. The bent tube *t* serves to draw off a sample of the gas in the jar *J*, and as it is drawn off, the water flows through the tube *l* from the beaker *o* outside, into the second vessel inside *i*. The tubes *t* and *l* are now clamped, and the apparatus exposed to bright light for 4 or 5 hours when a fresh sample of gas is drawn off and analysed. The water introduced absorbs some of the CO_2 and causes an error, which however is not so serious as to interfere with the results for demonstration purposes.

(52)   *Timiriazeff's Eudiometer*[1].

For the analysis of gas given off from water plants we use Timiriazeff's Micro-eudiometer arranged for the analysis of larger quantities of gas, e.g. for 0·5 c.c. instead of "a bubble no bigger than a pin's head."

The apparatus (shown in fig. 10) consists of three parts —the eudiometer *E*, the pipette *P*, and the carrier *C*.

The eudiometer is a tube of 5 mm. internal diameter graduated in 0·01 c.c.: the upper end is covered by a short (25 mm.) length of rubber tube through which passes a glass rod *R* serving as a piston. The lower end of *E* is enlarged into a small funnel *F* to facilitate the entrance of the gas to be analysed. The carrier *C* consists of a bell-shaped glass vessel 10 mm. in diameter, 20 mm. in height holding about 1 c.c. and attached to a glass rod *H* serving as a handle.

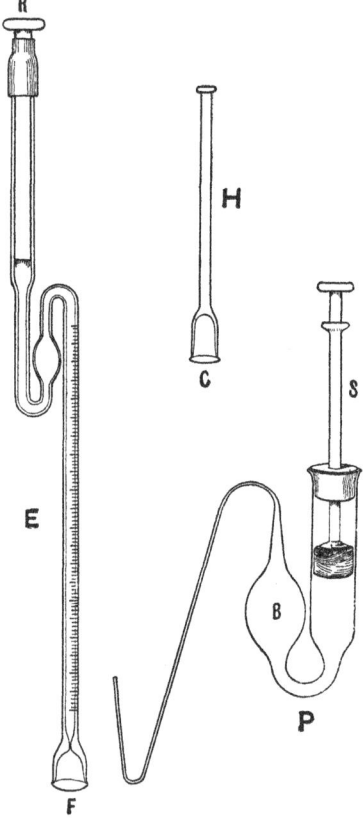

Fig. 10.   Exp. 52.

The carrier is filled with water and fixed so that the bubbles coming off from a water plant collect in it.   It is there transferred to the vessel[1] of water in which E stands, and the gas in C is poured up into the funnel F at the lower end of E. The funnel is connected with the graduated part of the tube by a capillary passage[2], so that the gas transferred from the carrier remains in the funnel, whence it is drawn into the eudiometer by pulling out the glass rod R working like the piston of a syringe. The gas having been measured by means of the graduations on E, the piston is pushed in and the bubble forced down into the funnel F.

The pipette P is dilated into a bulb at B and ends below in a bent capillary tube which can be inserted into the funnel of the eudiometer, when by drawing out the pistons of the syringe at the upper end of P, the gas is drawn into the pipette. For estimation of oxygen the pipette contains freshly made solution of potassium pyrogallate[3]; a second pipette filled with potash solution serves to estimate the CO_2. After two or three minutes, the gas is returned to the funnel of the eudiometer, drawn in by the eudiometer syringe, and once more measured. The difference between the first and second readings of the eudiometer gives the amount of oxygen absorbed by the pyrogallate, or of CO_2 by the KOH, as the case may be: the whole operation is so quickly done that no corrections for barometric or thermometric changes are necessary.

[1] The vessel should be of a dark colour so that the glass funnel at the lower end of E may be perfectly visible.

[2] This arrangement, which is the only essential point of difference between our apparatus and Timiriazeff's, is due to Mr F. F. Blackman.

[3] 0·4 gram pyrogallic acid in 20 c.c. of KOH (30—50 p.c. solution).

Since an apparatus of this kind will usually be made
in the laboratory it is worth noting that the bore and
length of stroke of the syringes, as well as the size of the
bulbs blown on the pipette and eudiometer must be
arranged so that the gas cannot be drawn beyond the
bulbs in either case.

(**53**) *Engelmann's bacterial method.*

This depends on the extreme sensitiveness of certain
bacteria to the presence or absence of free oxygen.   One
of the difficulties connected with the experiment is the
providing a sufficiently sensitive bacterium.   Pfeffer
recommends that a pea having been killed by boiling
shall be allowed to putrefy in 200 c.c. water; according
to Detmer a pure culture should be made of the bacteria
so obtained.   This, though no doubt advisable, is not
necessary.

It is best to begin with a study of the behaviour of
bacteria mounted simply under a cover-glass.   They will
be found to swarm round any air-bubbles which may be
included in the fluid under the cover-glass; and to collect
round the edges of the preparation, and in fact to seek out
sources of free oxygen.   If the preparations are sealed by
a coating of melted vaseline or wax-mixture painted
round the edge of the cover-slip, the bacteria ultimately
become sluggish and come to rest.   It is of this fact that
Engelmann's method takes advantage.   If a filament of
Spirogyra or the leaf of a submerged plant be included
with the sealed bacteria we have it in our power, by the
exposure of the preparation to light,—to produce free

oxygen. Thus all that is necessary is to place the preparation in the dark for two or three hours, then to expose it to light, and to watch the swarming of the bacteria round the green plant. The bacteria will be in violent movement within half a minute after exposure to light. If not kept long in the light they may be brought to rest by a quarter of an hour's darkness.

By means of Engelmann's Microspectral Objective it is possible to cast a spectrum on the filament of Spirogyra and to observe the distribution of the swarming bacteria in the different colours. We do not propose to enter into Engelmann's method of "successive observations," for which the student may consult Engelmann's papers in the *Botanische Zeitung* from 1881 onwards.

(**54**) *Diffusion.*

In connection with assimilation the diffusion of gas through the cuticularised epidermis should be studied. Detmer's method[1] may be used.

A pierced rubber cork is fitted over a glass tube (3 cm. diameter) so that the surface of the cork is flush with the upper rim of the tube. On the aperture in the cork a piece of fine wire-gauze is laid, and on this a leaf (e.g. that of *Platanus* or of *Nerium*) is placed with the stomatal surface uppermost, and firmly cemented with wax-mixture to the cork. The tube is filled with CO_2, and its lower end plunged into mercury. As the CO_2 diffuses out through the leaf, the mercury rises in the

---

[1] *Praktikum*, p. 107.

tube. The wire-gauze serves to prevent the leaf bulging inwards into the tube. The best method of filling the tube is by displacement of the air, which is allowed to leave the tube by a small gap purposely left uncemented between the leaf and the cork, and which can be closed when the air has been replaced by CO_2.

SECTION C. **Reactions of chlorophyll and of some other pigments.**

To study the simpler reactions of chlorophyll we extract the green colour of leaves by means of alcohol. The leaves[1] are boiled for a few minutes in water, roughly dried with filter-paper and placed in alcohol. The extraction must go on in the dark, because light has a destructive action on the colouring matters.

(**55**) *Separation by Benzol, etc.*

Place some of the alcoholic extract in a test-tube, dilute it with a few drops of distilled water; add benzol, shake the mixture, and allow it to settle. The benzol which floats above the alcohol is of a bright greenish blue, while the alcohol dissolves the yellow pigment which forms part of the alcoholic leaf-extract.

A similar separation may be effected by adding to the alcoholic extract :—

    (*a*)   Ether.
    (*b*)   Olive oil.

---

[1] Almost any leaves will serve the purpose : grass answers well.

(56) *Action of light*[1].

Fill three test-tubes with alcoholic leaf-extract, cork them and place *A* in sunlight, *B* in diffused light, *C* in the dark. After a few hours note the changes in colour. The solution which has been exposed to sunlight rapidly becomes brown or yellowish brown, while *C* is unchanged and *B* is intermediate in tint. In the absence of sunlight the effect may be shown by placing *A* close to the window, *B* in a dull corner, and *C* in the dark. Exposure for 24 hours is necessary. Chlorophyll solution may be compared with an alcoholic extract of etiolin, which is almost completely stable in light.

(57) *Aeration in connection with the action of light*[2].

Boil some of the alcoholic solution in a test-tube, so as to remove the air, cork it[3] and allow it to cool. Place it with an unboiled sample in bright diffused light, and note that the absence of oxygen delays the light effect.

(58) *Action of acid.*

Add a few drops of HCl to the alcoholic extract and note the appearance of a brownish tint; with excess of acid a muddy blue is produced owing to the precipitation of phyllocyanin and phylloxanthin[4].

[1] Sachs' *Bot. Zeitung*, 1864; *Physiologie* (French Trans.), 1868, p. 13; Wiesner, *Sitz. Wien. Akad.* vol. LXIX. 1874.

[2] N. J. C. Müller, *Pringsheim's Jahrb.* VII. p. 205.

[3] To cork a full test-tube, the best plan is to include a piece of thin wire between the cork and the glass; this makes a vent for the escape of the fluid, which closes when the wire is pulled out. To prevent the entrance of air the operation of corking should be *finished* with the test-tube inverted in a fluid, and the wire should be pulled out under the same conditions.

[4] See Schunck, *Annals of Botany*, III. p. 88.

(**59**) *Action of copper salts.*

By the addition of a little 10 % CuSO$_4$ solution[1] a copper compound is produced, which has the general appearance of chlorophyll, but differs notably in not being fluorescent[2]. To observe this point compare it with unaltered chlorophyll extract; fluorescence is most easily visible with a strong solution in a narrow test-tube.

(**60**) *Stability of the copper compound*[3].

Fill two test-tubes *A*, *B*, with the copper compound and two others *C*, *D*, with unaltered leaf-extract: place *A* and *C* in sunlight, *B* and *D* in the dark. After some hours note by comparison with *B* and *D* that the copper compound is not destroyed while *C* is affected.

(**61**) *Spectroscopic examination.*

To see the characteristic chlorophyll band *I* in the red, a small direct-vision spectroscope may be used: the solution may be in a test-tube, and ordinary daylight will suffice. In Detmer's *Praktikum*, p. 17, a convenient holder for test-tubes is figured and described. For the other bands direct sunlight is needed, the solution, which must be a weak one, should be placed in a parallel-sided vessel, and a more elaborate spectroscope should be used.

(**62**) *Other pigments (Anthocyan).*

The red varieties of *Ricinus* and *Amaranthus* may be

---

[1] Or of strong solution of copper acetate and strong HCl.
[2] Tschirch, *Deut. Bot. Ges.* Bd. v. 1887, p. 135.
[3] See Schunck, *Annals of Botany*, III. p. 94.

used. In the last-named the red colour can be obtained
by boiling a leaf in water, which takes out the coloured
cell sap, and leaves the leaf green. In the case of *Ricinus*
the red colour is destroyed by boiling. If these leaves are
partly immersed in boiling water, the parts which have
been heated reveal, almost at once, the chlorophyll. The
red colour may be restored to the fluid in which the leaf
has been heated, and also to the leaf itself by acid. The
explanation of the facts given by Molisch[1] is that as soon
as the leaf is killed, the strong alkalinity of the protoplasm
makes the anthocyan alkaline, when it is greenish or nearly
colourless. According to the same author the leaves, e.g.
those of *Amaranthus*, which do not lose their red colour
on being boiled, contain an acid cell sap which is not
entirely neutralised by the alkaline protoplasm, and therefore preserves the red colour of the anthocyan.

(**63**) *Floridece.*

In some species, at any rate, the colouring matter
reddens cold fresh water in which the sea-weeds are placed,
but the colour is destroyed by boiling. In *Polysiphonia* it
is not destroyed.

(**64**) *Brown sea-weeds.*

A portion of *Fucus* or *Laminaria* yields a brown colour
to water in which it is boiled—while the boiled thallus
shows a greenish colour and yields a green alcoholic
extract. But it is impossible as far as we have seen to
extract the whole of the colouring matters.

---

[1] *Botan. Zeitung*, 1889, p. 20.

SECTION D.   **Conditions necessary for production of chlorophyll.   Etiolation.   Sun- and shade-leaves.**

(**65**)   *Formation of chlorophyll*[1].

Seedlings of mustard (*Sinapis*) are grown in the dark[2] and are then placed in the morning in a good light close to the window, and the time necessary for the production a of distinct green colour is noted; some effect is visible after one hour.

Place similar etiolated plants in the darkest corner of the laboratory and when chlorophyll has been developed show, by an examination of the leaves with Sachs' test, that light too weak for assimilation is strong enough for chlorophyll-formation.

(**66**)   *Etiolin and light.*

The following point is of less importance.   Compare the colour of etiolated seedlings, which have been exposed to light for one or two hours but have not developed chlorophyll, with control specimens left in the dark.   They will be found to be of a darker yellow or orange colour. In this way Elfving[3] showed that light increases the formation of etiolin.

(**67**)   *Pinus.*

Light is not necessary for chlorophyll formation in

[1] Wiesner, *Die Entstehung d. Chlorophylls*.   Vienna, 1877.

[2] Etiolation proper can only be observed in parts of plants which have developed in the dark.   The already formed chlorophyll may become discoloured by starvation, but this is not etiolation.   Many leaves retain their green colour for a long time in darkness.

[3] Sachs' *Arbeiten*, II. p. 495.

certain Gymnosperms[1]. The seeds of various species of *Pinus* should be sown three weeks or a month before they are needed for demonstration. Let them be kept in the dark continuously and at a temperature of at least 15° C. Peas or beans should be grown with them to prove by their appearance that the cupboard is dark enough to etiolate ordinary plants.

(68) *Temperature*[2].

Sink an empty beaker in a larger one half filled with water, and keep the water at 30° or 31° C. by means of a thermostat. Etiolated plants such as seedlings of *Sinapis* or the epicotyls of beans are placed in the inner beaker, which is covered by a glass plate. A similar vessel contains control plants and is allowed to remain at a room temperature of about 15° C. After 2 or 3 hours a distinct difference in the greenness of the plants at 31° C. as compared with the control plants is perceptible. In one experiment with mustard seedlings, in which the control plants were kept at a temperature of 10° C., a distinct effect was perceptible in one hour.

(69) *Oxygen necessary for chlorophyll-formation.*

Germinate mustard in the dark and when the cotyledons are free from the seed-coat pass two or three plants under the rim of an inverted test-tube filled with water. They float[3] up to the top of the tube and are thus fully

[1] Sachs' *Physiologie* (French Trans.), p. 8.

[2] Sachs, *loc. cit.* p. 11.

[3] If the seedlings sink instead of floating, as sometimes happens, they may be allowed to lie at the bottom of a beaker of water.

exposed to light, but they do not become green; while control plants placed on wet filter-paper under a bell-jar soon develope chlorophyll. It is not necessary to use boiled water, the amount of air in ordinary spring water being insufficient for the respiration of land-plants.

(70)   *Seedlings in hydrogen.*

To demonstrate the fact in another way mustard seedlings may be placed in hydrogen. We use the L-shaped vessels recommended by Detmer[1]. The difference between the experimental seedlings and the control in air is clear after 24 hours. The vessel may be filled with hydrogen by displacement of water.

(71)   *Iron.*

The effect of iron salts in restoring a green colour to chlorotic[2] leaves, may be occasionally demonstrated on chance specimens. A chlorotic branch of Robinia in the Botanic Garden at Würzburg was restored to a healthy green by screwing a funnel into the tree close to the base of the branch, and pouring into it a solution of an iron salt[3].

In the absence of chance material, chlorotic plants must be produced by growing them, by the water culture method, without iron. It is best to grow some five or six iron-starved plants so as to have control plants and to make sure of material for several experiments. The addition of a few drops of a solution of iron chloride should lead to the development of chlorophyll, but ac-

[1] *Praktikum*, p. 26.   [2] See Sachs' *Arbeiten*, III. p. 433.
[3] Sachs' *Vorlesungen*, p. 343.

cording to our experience success is by no means certain. Another jar may be used for Gris'[1] experiment, which consists in painting a leaf with very dilute ferric chloride solution. Here again it is not easy to insure success.

(72) *Form of etiolated plants*[2].

For a thorough study of the changes of form and structure which accompany etiolation it would be necessary to grow a great variety of plants. The best for the purpose are plants produced from tubers or bulbs, or from large seeds full of reserve material, since here the effects of darkness in producing starvation do not complicate the result. Among Dicotyledons, *Dahlia, Helianthus tuberosus, Humulus lupulus*, and Beans (*Faba* and *Phaseolus*) may be grown. Among Monocotyledons, any of the cereals, *Narcissus* and *Crocus*.

In each case control plants of the same species must be grown in light. Compare the two sets as to development of leaf, measured in length and breadth; length and diameter of stem, and length of internode.

(73) *Sun- and shade-leaves.*

To see the remarkable structural characters described by Stahl[3], the leaves of the beech (*Fagus*) will serve. Transverse sections must be cut from leaves which have grown (1) in the fullest sunshine, and (2) in deep shade. The chief point to note is the difference in the palisade tissue.

[1] For an account of the experiments of Gris see Sachs' *Physiologie* (French Trans.), p. 159. Also Sachs' *Arbeiten*, III. p. 433, for Chlorosis.

[2] Sachs' *Bot. Zeitung*, 1863.

[3] *Bot. Zeitung*, 1880.

CHAPTER III.

SECTION A.   *Water-culture.*
SECTION B.   *Experiments on Fungi and on Drosera.*
SECTION C.   *Absorption and other functions of the root.*

SECTION A.   **Water-culture.**

(74)   *Method.*

To show what elements are necessary for the development of a green plant, and the relative proportions in which they are absorbed by its roots, the method of water-culture should be used, either alone or in combination with studies on the ash obtained by incinerating the plants cultivated.

Full directions for conducting water-culture experiments are given in Sachs' *Lectures*, Eng. ed. Lect. XVII. p. 283, and in Detmer's *Praktikum*, ch. I. pp. 1—6[1].

Although we have never succeeded in preventing the failure of a small proportion of such experiments, the liability to failure may be much diminished by careful attention to the following precautions.   The cylinders

---

[1] Compare also Acton, *Proc. Royal Soc.* Vol. XLVII. (1889), pp. 152–157.

used should not contain less than 500 c.c. of the solution[1] in an experiment and should therefore be of at least 700 c.c. capacity[2]. Every cylinder used should be carefully cleaned just before setting up the experiment. For this purpose the cylinders are thoroughly washed, and then rinsed out with strong commercial nitric acid which is removed by distilled water. They are then again rinsed out with a strong aqueous solution of mercuric chloride, and lastly with distilled water, which has been boiled for some time immediately before use, till portions of the wash-water give no trace of turbidity with a solution of silver nitrate.

[1] Sachs recommends the following :

Potassium nitrate	1·0 gram	
Sodium chloride	0·5	
Calcium sulphate	0·5	
Magnesium sulphate	0·5	
Calcium phosphate	0·5	
Water	1000 c.c.	

Pfeffer, *Physiologie*, Vol. I. p. 253, quotes from Knop the following :

Calcium nitrate	4 parts by weight		
Potassium nitrate	1	,,	,,
Magnesium sulphate (crystals)	1	,,	,,
Potassium phosphate	1	,,	,,

One part of the mixture of salts is dissolved in 50 parts of water : for use it is diluted to 2 or 3 per mille. A drop or two of iron chloride must be added to it as in the case of all normal nutrient solutions. Schimper (*Flora*, 1890, p. 220) gives a variety of useful formulæ.

[2] Wortmann (*Bot. Zeitung*, 1892, p. 643) recommends the use of very large vessels for water-culture. He states that in this way the respiration of the roots is better provided for, and that culture fluid remains at a lower and healthier temperature. According to Wortmann the plants flourish far better than in the ordinary small vessels, and moreover require practically no further attention when the culture has once been set up. He uses glass cylinders containing 26½ liters, which are supplied at a cost of 5 marks each by Messrs Ehrhardt and Metzger of Darmstadt.

The culture solution should be boiled rapidly for at least half-an-hour, the water which evaporates off being replaced from time to time with pure distilled water, and transferred to the cylinder as soon as it has cooled.

Two holes should be cut in the cork, one for the plant and one for a tube to admit air to the interior of the cylinder. For the latter purpose a short glass tube is inserted through the hole in the cork so that the ends project about 5 cm. beyond the upper and under surface; the upper open end is attached to a small bulb tube loosely packed with recently ignited asbestos which will exclude dust etc. but allow a circulation of gases. This tube is also useful for introducing fresh water, when required, without touching the plant, as it is only necessary to remove the bulb tube and afterwards replace it.

To fix the plant in position in the cork, soft asbestos which has been recently heated is preferable to cotton-wool, and the material should not project beyond the lower surface of the cork, as it is desirable to keep it as dry as possible, since 'damping off' at the 'collar,' from the attacks of fungi, is the commonest cause of failure in culture experiments[1]. For the same reason, only those plants should be selected for use which are uninjured at the 'collar,' and great care taken that no injury is inflicted at this part when fixing in position. When changing the plants into fresh cylinders the whole

[1] Out of fifty-six unsuccessful experiments where plants died within three weeks, more than thirty were attacked in this way; the plants were seedlings of *Epilobium hirsutum* and *Cheiranthus cheiri*.

cork should be taken out and put into the new cylinder, but if for any reason the asbestos around the collars should get damp it is better to take a fresh cork and to fix the plant again with dry material.

At the end of each week the plants should be changed into cylinders containing only pure distilled water and left in the same for three or four days, when they may be again placed in the culture solution, using for this purpose a fresh 500 c.c. of the solution put into the vessels with the same precautions as at first. The longer such cultures are continued, if the plants keep healthy, the more striking will be the results, but three weeks, during average summer weather, will be sufficient to demonstrate the facts illustrated in the selected experiments.

Pure chemicals should be used in making up culture solutions; the solutions do not keep well even in the dark and should be freshly made for each set of experiments. A useful rough rule for making up such solutions is to dissolve twice the weights of the solids, given in grams per liter, in an ordinary blue glass Winchester quart bottle, containing roughly 2 liters.

Water-plants cannot generally be recommended for accurate experiments extending over any considerable time, as we have found it much more difficult to grow them satisfactorily in culture solutions than to grow ordinary plants with the roots immersed.

Strong seedlings of any common green plants may be used; of the plants used by Açton (*loc. cit.*) the best were found to be *Epilobium hirsutum* and *Cheiranthus cheiri*.

In experiments where the time required is not very

long, shoots of plants with the cut end in the solution may be used; shoots of *Alisma plantago* and *Scrophularia aquatica* are good for this purpose, and when it is convenient to have a woody stem, branches of *Acer pseudoplatanus* or *Tilia europœa* answer well.

(**75**) *Potassium salts necessary.*

Take three plants, *A*, *B*, *C*, as nearly as possible of equal weight and equally developed. Dry *A* at 100° C. and determine its dry weight. Grow *B* in normal culture solution and *C* in a fluid containing the same salts as the normal solution but with an equivalent weight of sodium —instead of the potassium—salt. Continue the cultures for about three weeks, then take out the plants *B* and *C*, dry them at 100° and determine their dry weights. *B* should be considerably heavier than *C*.

To confirm the fact that the greater increase in weight shown by *B* is associated with the actual absorption of the potassium, *B* and *C* should be incinerated after weighing and the absolute amounts of K_2O in the whole ash of each determined.

Instructions for obtaining the ash and making an accurate estimation of the K_2O are given in Part II.

(**76**) *Phosphoric acid necessary.*

The same method is used as in the last experiment but a somewhat longer time will be required for satisfactory results. The solution which contains no salt of phosphoric acid may have the usual calcium phosphate replaced by an equivalent quantity of calcium nitrate.

Instructions for determining P_2O_5 in the ash are given in Part II.

In this as the preceding experiment it need scarcely be pointed out that it is much better to start five or six separate cultures under each set of conditions than to rely on one only. If all develope well, the mean result of the best three may be taken in each case.

(77) *Experiments with Lemna.*

Though as above stated water-plants are not generally to be recommended, yet we have found *Lemna minor* useful for purposes of demonstration. They grow rapidly, and their increase being principally in one plane is easily noticed at a glance. Moreover a rough numerical estimate of the amount of increase in a given time can be made by counting the fronds; thus in fig. 11 the culture *S*, which has about 21 fronds, consisted originally of six separate fronds, as shown in culture *W*.

We grow the *Lemna* in narrow cylinders containing 300 c.c. of fluid; if the cylinders are darkened by black cardboard covers the cultures keep reasonably free from algæ.

Fig. 11 gives the result of an experiment carried on in a greenhouse in the winter. Three jars *S*, *K*, *W*, were prepared, in each of which six fronds were placed. *S* contained 0·25 % Sachs' mixture of salts; *K* contained 0·25 % potassium nitrate, while *W* contained only distilled water, a drop of dialysed iron being added to each culture. The amount of increase is shown in the figure, the difference in root production as well as in the amount of frond is noticeable. In this and similar experiments

the *Lemna* died in a short time in distilled water;
whether this is due simply to starvation or to some other
cause we have not ascertained.

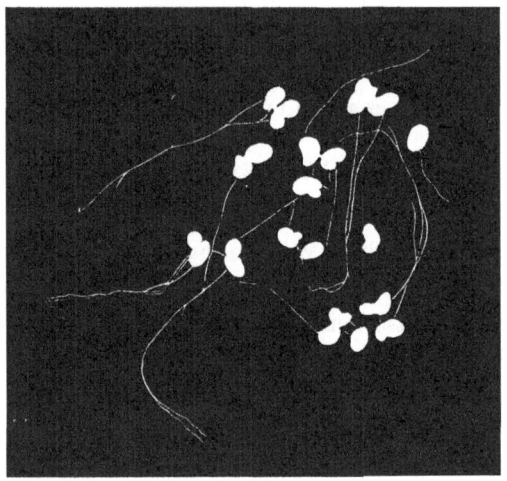

S

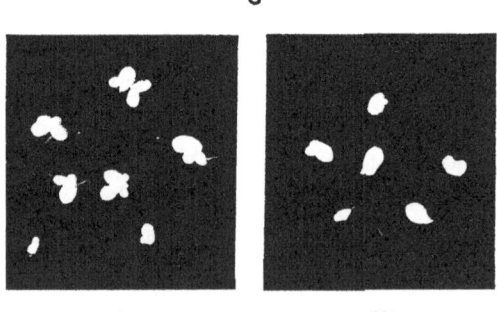

K   W

FIG. 11.   Exp. 77.   (Life-size.)

Fig. 12 gives the comparative result of culture in
Sachs' fluid (*S*) and in the same without phosphates

(P). Four or five weeks (in May) are necessary to give the result. Owing to an accident the figures do not show the strong growth of roots in (S)[1].

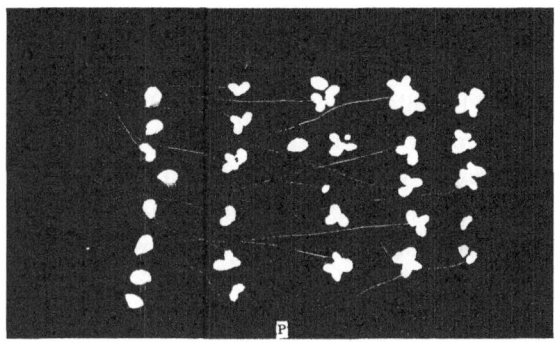

FIG. 12.   Exp. 77.   (Reduced.)

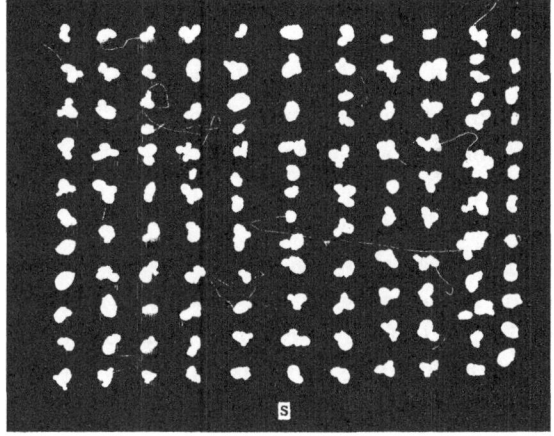

FIG. 12.   Exp. 77.   (Reduced.)

[1] In both cultures there were originally 6 plants each with 3 fronds.

(78)   *Calcium oxalate formation.*

The leaves of *Æsculus hippocastanum*, of *Acer negundo*, *Ulmus campestris*, and *Humulus lupulus* are according to Schimper[1] useful to demonstrate the fact that calcium oxalate accumulates in leaves with age. Young and old leaves of some of these species, having been rendered transparent by chloral hydrate, should be compared under the microscope. The method described in Chapter v. of detecting small amounts of calcium oxalate with the polariscope may be used, but will probably not be needed. The formation of the oxalate is connected with illumination: in *Æsculus*, as Schimper states, this is especially noticeable, leaves which have grown in full sunshine having far more crystals than leaves developed in the shade. The formation is also connected with the presence of chlorophyll. The comparison of a pure green and a white leaflet of a leaf of *Acer negundo* is, as Schimper states, especially instructive. In the white leaflets only a small amount of minute crystals occur. The variegated *Pelargonium* may also be used.

(79)   *Nitrate reaction.*

Schimper has shown that the appearance of calcium oxalate is connected with the decomposition of calcium nitrate in the leaf. The calcium being deposited as an oxalate while the nitric acid is assimilated. The disappearance of nitrate out of leaves shows therefore the same relation to light and to the presence or absence of chloro-

[1] *Botan. Zeitung*, 1888, p. 83.

phyll that he has shown to exist for the oxalate formation. The presence of nitrates is to be tested by the diphenyl-amin-sulphate test[1]; a not too thin section of a leaf or leaf-stalk is placed on a glass-slide and a drop of diphenyl-amin sulphate added; if nitrate is present a deep blue colour appears. Schimper recommends the leaves of the elder, *Sambucus nigra,* adding that the large leaves developed on the long spring shoots should be avoided, and that leaves developed in shade on short twigs should be employed. The cut leaves having been tested and found to contain nitrate are placed with their stalks in water and exposed to light. He describes an experiment in which the leaves lost the greater part of the nitrate in four or five days under these circumstances[2]. When a variegated elder is used for the experiment, the diminution of nitrate takes place in the green, not in the chlorotic parts. The importance of light was also shown in the case of *Taraxacum dens leonis, Aristolochia sipho* and some other plants by observing that after some weeks of sunny weather the sun-leaves gave no nitrate reaction while the shade-leaves showed a moderate or even strong reaction. We find that plants of *Pelargonium zonale,* grown in pots, give good results in a few days, in summer. One set should be exposed to bright light, the other set kept in deep shade.

[1] Molisch, *Deutsch. Bot. Gesellsch.* 1883. For the precautions necessary in drawing conclusions from observations based on this test see Zimmerman, *Botanische Mikrotechnik,* 1892, p. 49.

[2] According to our experience the lamina, not the leaf-stalk, of *Sambucus* should be tested.

SECTION B. **Nutrition of Fungi[1] and of Drosera.**

(80) *Method.*

Make the following[2] nutritive solution (N).

Dextrose	5 to 10·0 grams
Peptone	1 to 2·0
Ammonium nitrate	1·0
Potassium nitrate	0·5
Magnesium sulphate crystals	0·25
Potassium monophosphate	0·25
Calcium chloride	0·01
Pure water	100·0

Take 1000 c.c. of solution N and add to it 100 grams of pure gelatine (Coignet's gold label); sterilise in a flask plugged with cotton-wool, filter while hot and distribute into sterile plugged tubes: sterilise and preserve for use.

Expose a saucer of solution N to the air, until it is infected with one of the blue moulds:—*Penicillium* or *Aspergillus.* With a sterilised needle remove spores of the mould selected and shake up in a small flask of pure water: rapidly filter through a sterile funnel, plugged with cotton-wool to make the spores separate from one another. Add one drop, or more, according to the quantity of spores in the water, to a tube of the gelatine just liquefied, and pour it into a sterile glass dish. When set, put it aside at 20° C. in the dark; after 48 hours or so there will

[1] For the form of the instructions here given we are indebted to Professor Marshall Ward.

[2] Or any of the solutions given on p. 172 of Zopf, *Die Pilze.* Solution N is compiled from Elfving, *Studien über die Einwirkung des Lichts auf die Pilze*, 1890, p. 30.

probably be isolated pure cultures of the mould. Take spores from these with a sterile needle, and touch the nutrient gelatine of a series of the prepared tubes : this gives pure cultures of fungus for stock.

(81) *Various cultures.*

Prepare a series of small flasks (200 c.c.), plugged with cotton-wool and sterilised. To the flasks (A to E) add 50 c.c. of the following liquids :

    A. Pure distilled water.

    B. Solution N minus the dextrose.

    C. Solution N minus the peptone and nitrates.

    D. A 10 % solution of dextrose only.

    E. Solution N.

[N.B. These experiments need the greatest possible care to avoid any trace of impurity in the salts, water etc.]

Add to each flask one drop of pure water in which spores have been shaken, and separated by filtering through cotton-wool as described above, taking care that the drop contains only a few spores. If properly done each drop should contain about a dozen spores. Place the flasks in a temperature of 20° to 25° C., and compare the growths, which will be as follows :—

    A. No perceptible growth[1].

    B. Fair growth at first which soon, however, comes to an end.

[1] The miscroscope shows that the spores germinate, but the mycelium does not continue its growth.

C. Hardly perceptible growth which soon stops.

D. Fair growth at first, ceasing soon.

E. Standard growth, rapid and large.

If sufficient care is taken as to *absolute purity* (a difficult bit of manipulation[1]), it is possible to show, by leaving one out at a time, that each of the salts mentioned is necessary.

Also to show that, with *Penicillium*, magnesium sulphate can be replaced by magnesium sulphite or hyposulphite, but not by some other sulphur compounds.

K by rubidium or cæsium, but not by Na, Li, Ba, Sr, Ca, Mg.

Ca by Mg, Ba, or Sr, but not by K or Na[2].

(82) *Puccinia.*

Obtain teleutospores of *Puccinia graminis* which have wintered on the straw of wheat or *Triticum repens*, and sow in February—April in, (A) water, (B) nutritive solution, and keep at 10—15° C. in the dark.

Both will germinate, and even proceed to develope the sporidia, but these die off eventually.

Their further developement can only be got by infecting young leaves of the barberry (*Berberis vulgaris*).

The same thing is true of other *Uredineæ*.

(83) *Hanging-drop cultures.*

A damp-culture cell is to be prepared as follows[3]. A

---

[1] Owing to the cotton-wool, dust, glass, water &c. rather than the chemicals themselves.

[2] See Nägeli, *Ernährung der niederen Pilze.*

[3] H. Marshall Ward, *Philosophical Transactions*, 1892, B, p. 130.

deep glass ring is placed on a broad glass slide and a drop of previously sterilised olive oil allowed to run in, or melted paraffin may be run in, in the same way while the slide and ring are hot; this cements the ring to the slide, while a cover-slip placed on the ring like a roof supports the hanging drop. Or a chamber may be made as described and figured by Marshall Ward, *loc. cit.* p. 131, which is especially useful where it is desired to control the nature of the atmosphere to which the drop is exposed.

Everything being sterilised and the cell ready, take a clean cover-slip, heat it between two sheets of talc over a flame, and allow it to cool. Then, with forceps, place the cover-slip on any convenient support, and with a platinum needle place a drop on the centre. The drop is got thus:—

Infect a tube (gelatine or fluid medium) with a drop of water containing spores, and shake thoroughly. Hold a platinum needle in a flame, and let it cool; dip it into the infected medium and place the drop on the cover-slip. Then rapidly invert the latter, and cement it to the cell with gelatine, or with oil, or paraffin.

The drop should contain *one* spore, and trials have to be made to insure this. In gelatine media, the student can work with two to five spores if well isolated.

All the foregoing experiments can be repeated with drop-cultures.

(**84**) *Germination.*

Place a culture containing *one* spore in focus under the microscope. Record the temperature, and fix the spore

under the eye-piece micrometer, and cover the whole with
a darkened bell-jar. Examine the preparation from time
to time, and note the stages of germination. Measure the
germinal filaments, mycelial branches &c., and plot out
the rate of growth on sectional paper.

(**85**) *Drosera: digestion of white of egg*[1].

Drosera may be grown in wet moss in soup-plates : the
moss should be running with water which may advanta-
geously be changed every few days. *Drosera* cannot be
successfully cultivated in large towns. For the experiments fresh young leaves having good drops of secretion
on their tentacles should be selected. From the white of
a hard-boiled egg cut cubes of which the side measures
about a millimeter in length : place two of such cubes on
each of several leaves, and at the same time put other
cubes on the wet moss to serve as a control. They should
be examined in 24 hours and again after a further interval
of 24 hours. It will be seen that the egg on the *Drosera*
shows a distinct rounding at the angles of the cubes,
which are afterwards converted into spheres surrounded
by zones of transparent fluid. Still later the spheres
generally disappear and nothing but a small quantity of
viscid fluid is left.

(**86**) *Drosera: benefit derived by feeding*[2].

The plants are, as in exp. 85, to be grown in soup-
plates, each of which holds from 20 to 30 plants. Each

---

[1] C. Darwin, *Insectivorous Plants*, p. 93.
[2] F. Darwin, *Linnean Society's Journal*, Vol. XVII. For references to
other similiar experiments see *Insectivorous Plants*, 2nd Edit. 1888, p. 15.

plate must be divided in two by a thin wooden partition, this serves to mark off those plants which are to be fed from those which are to receive no food. Roast meat is cut across the grain into thin slices and the fibre teazed and cut into fragments so small that 15 together weigh 2 centigrams. A given leaf should not receive more than two of these particles at a time; they may be placed on the glands of separate tentacles : the feeding may be repeated every four or five days. The plants should be grown under wooden frames covered with fine netting (mesh 1·5 mm.) to exclude insects. The fed plants soon begin to look clearly greener and more vigorous than the unfed ones. To get a good result the experiment should be begun in May or June and continued to the middle of August. The number and height of the flower scapes, the number and weight of capsules, the number of seeds per capsule, &c. should be compared. Or the plants may be carefully washed and dissected out of the moss and the dry weight per plant of the fed and starved specimens compared.

Section C. **Roots**.

(87) *De Saussure's experiment*[1].

When plants are placed in solutions of various salts they do not, except under certain conditions, absorb the water and salt in the same proportion. De Saussure, using solutions that were not very dilute, found that the plant absorbed relatively less salt than might have been

[1] De Saussure, *Recherches chimiques*, 1804, p. 247.

expected. This condition of things is sometimes spoken of as absorption according to De Saussure's law, and although it is well known to be only a special case, the fact itself is worth confirming. In our experiments we proceeded as follows.

A bunch of rooted water-cresses (*Nasturtium officinale*) was taken up, washed and placed in distilled water for three days to allow the roots to recover from their injuries. They were then placed in a beaker containing 700 c.c. of a solution made by dissolving 1 part of potassium chloride in 1000 parts of water. They were left in the fluid for 8 days, by which time only 260 c.c. of solution were left in the beaker. This was analysed volumetrically, by titrating with decinormal silver nitrate, using potassium chromate as indicator. If the salt and the water had been absorbed in the same proportion the remaining solution should have still contained 0·1 p.c., i.e. 0·26 grams; in other words, the plant should have absorbed 0·44 grams. It was found however that less than this had been taken up, and that $\frac{5}{?}$, i.e. 0·5 grams, of the original potassium chloride instead of 0·26 grams were still present. Other salts give various different coefficients for this same strength of solution. If sufficiently dilute solutions be made use of, it has been found that, in contrast, relatively more salt than water is absorbed and the remaining portion of the liquid contains less than the due proportion of the original salt.

(88)   *Root pressure.*

Root pressure can be easily observed in young plants of

Phaseolus. An india-rubber tube T (Fig. 13) is tied on the cut stump, S, of the plant and is filled with water: a capillary glass tube G is tied into the tube, leaving about six centimeters of rubber tube full of water between the stump and the bottom of the glass tube. The glass tube is now fixed in a clip and after a time drops of water fall from the end E. To get an idea of the rate of flow it is only necessary to gently pinch the rubber tube so as to press the fluid out, and to absorb it with filter-paper held

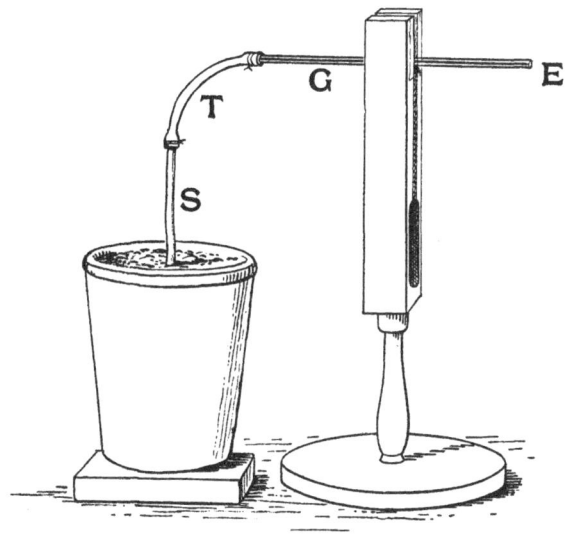

Fig. 13. Exp. 88.

at E.[1] When the rubber tube is released and therefore allowed to expand, a column of air is drawn into the tube

[1] The tube G, E, should be bent at right angles close to E, so that the terminal centimeter points vertically downwards.

and serves as an index of the rate of flow as it travels up the tube, which should be graduated.

By watering the earth with warm water a greatly accelerated rate of flow is obtained, but whether it is due to increased root pressure or to the expansion of air in the tissues is not easy to say.

(89)   *Root pressure.*

To demonstrate the force of root pressure a striking method is that used by Mr Gardiner in his lectures.   He uses a plant of *Sparmannia africana* growing in a large pot.   The stump is attached by rubber tubing to a poto-meter tube[1] filled with a solution of nigrosin in water ; to one arm of the potometer a vertical glass tube, a few mm. in diameter and several feet in length, is attached ; the other arm of the potometer is closed with a cork.   The nigrosin seems to have no bad effect on the plant and makes the rising column of fluid easily visible.   If the tube is supported against a wall it can be elongated by fresh lengths of glass tubing and thus a column of 8 or 10 feet can easily be shown.

(89 A)   *Root pressure.*

The classical method of observing root pressure is that described and figured by Sachs in his *Physiologie* (Fr. Trans.), p. 223, of which the following (Fig. 14) is a modification.   A **T** tube (*T*) having one arm *B* bent so as to be parallel to the two others, is tied into a piece of

---

[1] The arrangement is similar to that figured in Sachs' *Vorlesungen*, p . 328, fig. 211.

pressure tube which is also tied to the plant[1]. The arm B
passes through a rubber cork firmly tied into a wide-

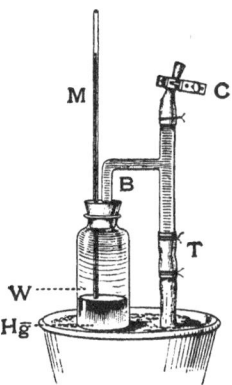

Fig. 14.   Exp. 89 A.

mouthed (stoppered) bottle, in the bottom of which is half
an inch of mercury, Hg: the tube M, which serves for
manometer readings, fits tightly into a hole in the cork
and reaches the bottom of the bottle. Water W is now
poured in at C so that the bottle and the arm T are filled.
At first the plant will usually absorb water, so that C
should be left open until the rise begins, when it may be
filled up and closed by means of a clamp. The mercury
will rise to a considerable height and will show diurnal
variations about its mean position.

(90)   *Moll's Experiment.*

Various kinds of plants, when placed under a bell-
jar standing in dishes of water, will give evidence of root

---

[1] The two ligatures at T are placed closer together than as represented in fig. 14.

pressure by the drops of water exuding from the leaves.
Root pressure may as Moll has shown[1] be replaced by that
of a column of mercury.   The branch or leaf-stalk, as the
case may be, is fixed air-tight into the short arm of a U
tube filled with water, and mercury is then poured into the
long arm until about 20 cm. pressure is obtained.   The
whole is then covered with a bell-jar standing in water,
and after a time drops of fluid are found hanging to the
leaves.   We found that with 25 cm. of mercury the drops
appear very rapidly on the leaves of the Balsam (*Impatiens
balsamina*).   Moll also recommends *Begonia* and *Phaseolus*:
in the last named the fluid is, as Moll says, found on the
lower surface of the leaf.

(**91**)   *Absorption by means of dead roots.*

Several observers[2] have shown that transpiring plants
can absorb water from the soil even after the roots are
dead.   We have confirmed the fact on pot-plants of
Helianthus tuberosus.   A thermometer having been forced
into the earth, the flower-pot is immersed in water so
hot that the soil is kept at a temperature of 60°—65° C.
for two hours.   In spite of this violent treatment the
leaves remain turgescent for several days, whereas control-
plants, shaken out of their pots and freed from soil, rapidly
wither.

[1] *Bot. Zeitung*, 1880, p. 49.   References are given to Sachs' *Lehrbuch*,
1874, p. 660, and de Bary, *Bot. Zeitung*, 1869, p. 883, for similar results.
[2] Strasburger, *Leitungsbahnen*, 1891, p. 849, where references to
earlier experiments are given.

CHAPTER IV.

TRANSPIRATION.

SECTION A. *Absorption of water by transpiring plants.*
SECTION B. *Loss of weight due to transpiration.*
SECTION C. *Stomata, Bloom, Lenticels.*

SECTION A. **Absorption.**

(92) *Potometer*[1].

In the first series of experiments (Section A) the rate of absorption of water by transpiring plants under varying circumstances is to be observed. This may be done with the potometer shown in fig. 15. Of the three openings of the potometer, A and B are closed by rubber corks; that in B is perforated by a capillary tube of about 0·3 mm. bore: the tube should just project beyond the cork on the inside and should have a total length of about 20 cm. The end A is closed by an unperforated cork, while to C is fitted about 8 cm. of rubber tubing, of which 4 cm. project beyond the end of the tube. The cork B should first be fitted in, then fill the potometer with water and

[1] F. Darwin and R. Phillips, *Cambridge Philosoph. Society*, Vol. v. 1886.

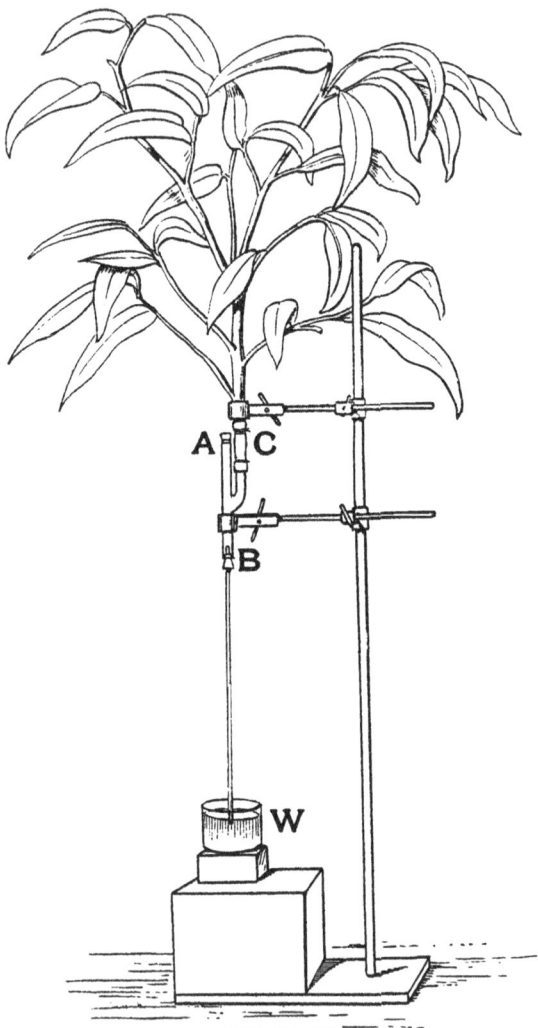

Fig. 15.   Exp. 92.

force the branch[1] into the rubber tube C, as far as it will
go. The joint between the rubber tube and the branch
must be secured by tying; for this it is best to use strong
uncovered elastic thread, which must be stretched while
it is being wrapped round the tube, and can be secured
by a simple tie, a complete knot being unnecessary. The
rubber tube may be secured to the glass tube with wire.
Turn the potometer upside down so that any air in C
may rise and collect at A, and before corking A, fill it to
the brim with water. Support the potometer on a firm
retort-stand and fix the plant to the same stand to avoid
any possible movement between the plant and the instrument. The end of the capillary tube dips into a
small vessel of water W supported on two blocks, of which
the upper one is small enough to be conveniently seized
in one hand, and of such a height that when it is removed
the capillary tube no longer dips in the water. When
this is done, (if the plant is absorbing vigorously) a
column of air will be sucked[2] in at the lower end of the

[1] If herbaceous plants, or woody plants with delicate leaves, are used
for transpiration experiments it is necessary to cut them from the
parent plant under water, to prevent the entrance of air, which rushes in
to satisfy the negative pressure. It is generally possible to force a branch
below the water in a basin held by an assistant, and divide it with a
strong pair of gardeners' shears. In the case of herbaceous plants the
process is made easier if a couple of sharp bends are made in the stem, a
proceeding which need not admit air, and has no effect in the transpiration current in the plant. For all research-work connected with the
transmission of water the specimens should be thus treated, but for the
following experiments, in which laurel or Portugal laurel (*Prunus laurocerasus* and *lusitanica*) are used, cutting under water is not necessary.

[2] To hasten the entrance of the air column, it is best to absorb, with
a piece of filter-paper, the water hanging at the end of the tube.

tube. As soon as this appears the small block is replaced, the tube once more dips into the water, and a bubble of air included in it travels up, and serves as an index of the rate of absorption. The bubble must be of uniform size in successive readings, because other things being equal a long and short bubble travel at different rates. To insure this, mark the tube with a file at 5 or 6 mm. from its end, and replace the vessel W when the air column has reached the file-mark. The movement of the air-bubble is timed from a mark on the tube : the upper limit of its course is the upper end of the capillary tube, the moment of its impinging against the water in the potometer at B being easily visible. It is for this reason that the tube projects above the cork, for otherwise a convenient place of ending for the course of the bubble would not be visible. The starting point of the course should be at least 4 cm. above the file-mark, so that the bubble may settle down to a uniform pace before the course begins ; and because time is needed for the observer to put down the block and the small vessel of water, and take up the stop-watch.

In this way numerous readings can be taken in a short time ; the air admitted collects at A, and can be removed after a time ; it is obvious if the branch were placed in A and the cork in C, that the admitted air might diminish the surface of contact between the water and the absorbing surface of the branch.

It is well to make graphic representations of one or more of the following experiments, using the reciprocals of the stop-watch readings, which will be proportional to the rates of absorption[1]. The actual volume of water absorbed per unit of time may be obtained by calculation from the length and diameter of the capillary tube. Or by replacing the plant by a siphon delivering a known weight of water per hour[2].

(93) *Kohl's method [slightly modified].*

This apparatus, like the potometer, is a modification of Sachs' instrument[3]. Here the index is not a bubble of air, but the column of air which travels onwards as the water is absorbed. We use a tube *abc* fitted up as

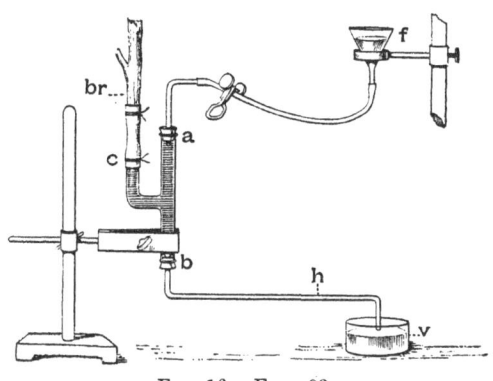

FIG. 16.   Exp. 93.

shown in fig. 16.   The end, *c*, as before, takes the branch *br*; *a* takes a tube connected by a rubber tube with a funnel *f*, and closed by a clamp; *b* takes a glass tube *h*, which is coarser than that used in exp. 92, and also

[1] Reciprocals may be got from published mathematical tables.
[2] See Darwin and Phillips, *loc. cit.*
[3] *Physiologie* (French Translation), p. 246.

longer[1].   When no readings are being taken the bent free
end of h dips into the vessel of water v; when v is removed
a column of air enters and travels along h, its rate of
movement being noted by timing it over intervals of 5 or
10 mm.   For this purpose h may be graduated, or a
millimeter scale may be set up behind it.   When the
column of air has nearly reached b it can be brought to
the zero of the scale by opening the clip and allowing the
water from the funnel f to drive it back along h.

(94)   *Sunshine.*

Take a branch of Portugal laurel (*Prunus lusitanica*)
which has been cut and placed in water for at least 6
hours.   This precaution is necessary to satisfy the negative
pressure in the branch; if this is not effected, variations in
the rate of absorption will by no means represent variations
in rate of transpiration.   Fit up the branch in the potometer[2] and take readings until the rate of absorption is
fairly constant.   Then place the plant in sunshine and
observe the increased rate of absorption, and finally replace
it in shade.

(95)   *Wind.*

When the rate is once more steady, open a door
and a window so that the plant is exposed to a draught.
The rate of absorption may easily increase by 50 per cent.

---

[1] The bore of the tube must be varied according to the size and
absorbing power of the specimen.   In the figure, h is represented shorter
than it usually is.

[2] Kohl's apparatus will answer for any of the experiments for which
the potometer is recommended, and *vice versa*.

(**96**) *Light and darkness.*

Kohl[1] has succeeded in demonstrating the effect of light on transpiration by a method which in his hands gives excellent results. Variations in transpiration are estimated by changes in the amount of water absorbed as in experiment 93. To avoid the difficulty of keeping the hygrometric condition of the air constant during the alternation of light and darkness, Kohl encloses the plant in a large bell-jar resting on a ground-glass plate. The leaves and stem of the plant are within the bell while the apparatus for recording the rate of absorption is below the glass plate and free to be manipulated. By means of an aspirator a current of dry air is drawn through the bell-jar, and the plant can be darkened by a cardboard cylinder slipped over the bell-jar without affecting the hygrometric condition of the air within. Kohl's curves show that the effect follows a change from darkness to light very quickly, and that the depression in the rate of absorption produced by darkness occurs still more rapidly. Plants of *Nicotiana tabacum* growing with their roots in water seem to give the best results.

We have noticed a source of error in this method which seems to be of importance. If the aspirator runs too quickly it is possible to produce diminished air-pressure within the bell-jar and thus produce an increase in the rate of absorption. We found it necessary to attach a mercury manometer to the apparatus in order that this source of error may be avoided. We have not

[1] *Die Transpiration der Pflanzen*, 1886, p. 61.

however had much experience of the apparatus and have never thoroughly tested it.

If the suction tube from an air-pump is fitted to the bell-jar and a good sized tube[1] for the admission of air is provided, it is easy to keep the air in the bell-jar dry enough to keep up a good transpiration current. This is an easier apparatus to fit up, because no air-tight joints are needed, but it is obviously faulty because the air in the bell-jar varies in dryness with the air of the laboratory. Still, from some few trials, we are inclined to think that the effect of light and darkness may be demonstrated in this way.

(**97**) *Negative pressure.*

The object of this experiment is to prove the need of the precaution mentioned under experiment 94, viz. leaving the cut end of the branch under water for some hours before using it. Cut a branch from a Portugal laurel (*Prunus lusitanica*), fit it at once to Kohl's apparatus, and take a series of readings. The absorption will be found to be quick at first, and then to become slower.

(**98**) *Negative pressure*[2].

Fit up a laurel branch in a potometer and allow it to remain for a day or so. Having taken a series of readings take out the branch and shave a few millimeters off the cut end, which will have become dirty: the partial

---

[1] A curved rubber tube which excludes light.
[2] See von Höhnel, *Botan. Zeitung*, 1879, p. 318.

stoppage of the vessels, which gives the dark tint, produces a slowing of the current, and when it is removed
the readings of the stop-watch show an immediate
increase in rate. When the readings are fairly steady
again, repeat the removal of a shaving of wood to prove
that this *per se* has no special effect.

(99) *Negative pressure.*

A similar result may be got more quickly by filling the
potometer with an emulsion of skim-milk diluted with three
times its volume of water. The slowing of the current,
and the recovery when the blocked ends of the vessels are
cut off, may thus be seen within a short space of time.

(100) *Negative pressure.*

The fact of the existence of negative pressure may
best be demonstrated by von Höhnel's method[1]. If a
strongly transpiring stem or branch be cut under a watery
solution of eosin, immediately removed and examined,
the red fluid will be found to have rushed into the
vessels to a considerable height. We find that " Solomon's
seal " (*Polygonatum multiflorum*) answers well. Or plants
of *Helianthus tuberosus* grown in pots and left without
water until they are on the point of withering. In winter,
seedlings of *Vicia faba* pulled out of the loose sawdust in
which they have grown, and allowed to lie on the table
until nearly withered, show good injection.

---

[1] *Pringsheim's Jahrbücher*, xii. 1879, p. 47 ; also *Botan. Zeitung*, 1879,
p. 297.

(101)   *Permeability of cell membranes*[1].

Negative pressure depends on the fact that although the walls of the xylem elements are extremely permeable to water, they present a great resistance to the passage of air—as long as they are wet.   It is therefore of importance to show that wet wood does not easily allow air to pass while the same wood dried is easily permeable.

The only important point about the experiment is to make sure that air emerging under pressure from dry coniferous wood is really passing through the walls of the tracheids and not escaping from the protoxylem or from flaws and cracks in the branch.   We find that the wood of *Pinus sylvestris* is decidedly better than that of *Taxus*. We use branches about a centimeter in diameter cut into bits of 2 to 4 cm. in length.   The branch should be cut and brought into the laboratory without being placed in water.   We have not generally used it until 2 or 3 hours have elapsed, but it is probably unnecessary to wait so long.   The most effective way of blocking the vessels of the protoxylem is to gouge a small cavity at the centre of the branch, which is afterwards filled up with modelling wax melted in with a hot wire.   If this is done at both ends of the bit of wood the vessels will be sufficiently closed.

Another plan is to turn cylinders of 1 cm. in diameter from the splint wood of *Pinus,* or of the silver fir (*Abies pectinata*), which Strasburger especially recommends.   The wood should be placed, while fresh, in alcohol, and if the

[1]   von Höhnel, *Pringsheim's Jahrbücher*, xii. 1879, p. 63 ; Pfeffer, *Physiologie*, i. p. 86 ; Strasburger, *Leitungsbahnen*, p. 729.

spirit is allowed to evaporate slowly, the wood (according to Strasburger[1]) is in a more suitable condition for experiment than if it is allowed to dry without being treated with alcohol. We have not compared the two, but the silver-fir wood dried after prolonged soaking in alcohol certainly answers well.

The three following observations must be made:

(i) Take a turned cylinder of dry wood, or a piece of *Pinus sylvestris* treated as above described (which answers perfectly and is more easily prepared), and attach it by a strong rubber tube to the short arm of a U tube. By pouring mercury into the long arm it will be found easy to force air through under a pressure of about 10 cm.; to make this obvious paint the upper surface of the wood with olive oil in which the escaping bubbles are visible. The surface ought to foam all round the circumference of the section. A chain of bubbles coming from a single spot suggests a faulty bit of wood.

(ii) A similar cylinder which has been thoroughly soaked in water is fitted into the U tube: the wet cell-membranes will be found to be extremely, though not absolutely, impermeable to air.

(iii) If the U tube is now filled with water it will be found that a very slight pressure of water forces water through.

(**102**) *Oozing of water from the lower end of wood.*

The experience gained in experiment 101 enables us

---

[1] *Leitungsbahnen*, p. 734.

to understand the following experiment[1]. A piece of a yew (*Taxus*) branch (4 or 5 cm. in length) is placed in water until thoroughly soaked. When removed from the water and held with its axis vertical no water escapes from the lower surface (although water is contained in the tracheids) because if water is to escape air must enter, and air does not easily pass wet membranes. If however a drop of water is added (e.g. with a wet paint-brush) to the upper surface, water immediately oozes out below.

(103) *Permeability of splint wood.*

A modification of the experiment may be used to illustrate the fact that the water travels in the splint-wood. A piece of yew branch (5 or 6 cm.) is fitted to a rubber tube of 50 or 60 cm. in length. The tube is filled with water and closed below by a clip. If the wood is held vertically with tube hanging straight down, the upper surface of the wood, which must be cut smooth, is dry. If the closed end of the tube is raised until it is slightly above the top of the branch, the surface of the young wood is seen to blush or change colour, even before the water can be seen to actually ooze from it.

(104) *Recovery of a flaccid shoot.*

De Vries has shown[2] that when a shoot is cut in the air it frequently withers after it has been placed in water. This has usually been explained as being due to the air rushing in under negative pressure and filling the vessels.

---

[1] See Sachs in his *Arbeiten*, II. p. 296. Also Godlewski, *Pringsheim's Jahrbücher*, xv.

[2] Sachs' *Arbeiten*, I. p. 287.

Whatever the explanation may be, it is interesting to note that a shoot which has been rendered flaccid, by being cut in the air and allowed to partially wither, can be rapidly restored to turgescence by forcing water into its vessels under pressure[1]. The cut end of such a withered shoot is attached by a rubber tube to the short arm of a U tube containing water. The position of the end of the shoot, which droops flaccidly over, is noted on a vertical scale, and mercury is poured into the long arm. Under the pressure of about 10 cm. mercury, the plant recovers and the end of the shoot can be traced with the naked eye rapidly travelling up the scale.

(**105**) *Sachs' emulsion experiment*[2].

To illustrate the fact that the pits of coniferous wood are closed and that the water-current must therefore filter through them, the following experiment is useful.

Prepare an emulsion of vermilion by adding a few paint-brushes full of good colour to a beaker of water, and filtering it through coarse filter-paper. Take a piece of yew 6 or 7 cm. in length and attach it by a rubber tube to the lower end of a glass tube a meter in length held vertically in a clamp. Fill the tube with emulsion and observe that colourless water drips from the lower end of the wood. After an hour or so remove the wood: note the red colour of the young wood due to injection of the cut tracheids with vermilion. Cut a shaving off the surface to show that the colour only extends to a small depth.

[1] Sachs, *Text-book of Botany*, Edition II. p. 683.
[2] Sachs' *Arbeiten*, II. p. 299.

An interesting modification of the experiment is to plunge the cut ends of transpiring branches into emulsion : in this way the distribution of the transpiration current in the cross section may be studied in various plants. Diluted skim-milk stained black with osmic acid makes a good emulsion.

(**106**)   *Injection of vessels*[1].

Cut two similar branches of Portugal laurel, *Prunus lusitanica*, place one in water, the other in melted cocoa-butter (or gelatine), into which it must dip as deeply as the vessel allows.   After an hour, during which the cocoa-butter is kept melted, take the branches out of the fluids, and let them lie on the table till the cocoa-butter is quite cold : cut fresh surfaces to both and place them in watery solution of eosin.   After an hour or two the progress of the eosin may be compared by cutting off shavings of bark at various heights in the two branches.

(**107**)   *Compression.*

To prove that the transpiration current travels in the vascular cavities, it may be shown that squeezing the tissues in a vice checks the upward stream of water[2].

Cut, *under water*, a leafy stem of *Helianthus tuberosus*, and fit it to Kohl's apparatus.   A plant with well-formed wood must be chosen,—young stems are too brittle. Branches of bramble (*Rubus fruticosus*) also serve, but are awkward to work with ; in winter, last summer's shoots of

[1] Elfving, *Bot. Zeitung*, 1882; Scheit, *Bot. Zeitung*, 1884; Errera, *Soc. R. de Bot. de Belgique*, Vol. xxv. Part ii. 1886.

[2] F. Darwin and R. Phillips, *Cambridge Philosoph. Soc.* 1886.

ivy (*Hedera helix*) answer fairly well. Do not attempt to compress the whole stem but cut away half of it before applying the vice. The exposed surface may for greater security be rubbed with lard to prevent air leaking into the vessels exposed ; in any case the part selected for compression must be as far as convenient from the cut end, so as to avoid the chance of air being sucked back into the apparatus. The vice should be a light one, e.g. a watchmaker's vice, so that it may support itself when it is screwed on to the stem.

When the rate of absorption is steady, compression may be applied : it will be found necessary to screw the vice with great force so that the compressed tissues are squeezed to a mere plate. If the compression has been continued for some time, and the vice is then unscrewed, it will be noted that the absorption is very rapid and that it soon slows down. This shows that negative pressure rises during compression and falls when water is allowed freely to enter the vessels.

(108) *Incisions.*

In some trees it is obvious that the amount of wood in the transverse section is far greater than is absolutely needed to carry the transpiration current. Fit a branch of yew[1] (*Taxus*) to the potometer, and take a few readings, then saw it half through and read again. The rate of absorption will be unaltered, and the branch may indeed be almost severed before the rate of absorption is seriously

[1] In the case of yew it is better to remove the bark (at any rate in the spring) because it is easily detached from the wood, and this makes it difficult to slip on the rubber tube.

depressed. The branch must be firmly supported in two places and the incision made between them, otherwise the weight of the branch will break the thin bridge of wood which is ultimately left.

When a slowing of the current has been clearly produced, divide the bridge[1] and compare its area with that of the rest of the splint wood of the branch.

(**109**)   *Cross-cuts*[2].

Take a branch of *Prunus lusitanica* or of *laurocerasus* which has stood some hours in water; fit it up in the potometer and, as in experiment 108, support the branch in two places, so that it may not break when incisions are made. Having taken a few readings, saw the branch half through at a spot between the two supported points, and 10—15 cm. from the cut end. To see clearly how far the saw-cut penetrates, and on which side of the branch it lies, it is advisable to push a square piece of cardboard (for instance a post-card) into the cut. This serves as a guide in making the next cut, which must be exactly opposite incision (i), and 2 cm. above it. It must be slightly deeper than incision (i), so as to overlap it; it is easy to make sure of this if a second card is placed in the second incision: the edges of the cards should be parallel and should slightly overlap each other. The points to note are that cut (ii) depresses the rate of absorption very much more than cut (i), and that after the fall, a rise in absorption-rate comes on.

[1] The bridge should be cut with a knife, the smooth surface so produced makes the area of the bridge easily perceptible.

[2] Dufour, Sachs' *Arbeiten*, III.; also F. Darwin and R. Phillips, *loc. cit.*

(**110**) *Course shown by eosin solution*[1].

Remove from the potometer the branch used in experiment 109, and place the cut end, for one or two hours, in strong watery solution of eosin, taking off the bark of the part in which the saw-cuts were made, leaving, however, 4 or 5 cm. at the base unpeeled. The course of the fluid as it passes up the stem is now traced by the eosin, the manner in which the colour spreads at the doubly-cut region being of course the chief point to be noticed. The reason for leaving the bark on the terminal 4 cm. (which is not an essential precaution) is simply to ensure that any superficial rise of fluid shall take place on the bark instead of on the wood.

(**111**) *Air-pump.*

Cut a branch of Portugal laurel (*Prunus lusitanica*) of the same size as that used in experiment 109, and select one having a part of about 25 cm. in length bare of side branches; leave it in water for some hours, then cut off the 25 cm. and attach one end of it to a potometer, and the other to a water air-pump. When the air-pump is in action water will be sucked through the branch out of the potometer and readings can be taken with a stop-watch. Adjust the suction of the pump so that the readings of the potometer are roughly the same as those obtained in experiment 109. Now make the two overlapping saw-cuts as explained under experiment 109 and note the result. The point of interest is that here there is no recovery after the depression in rate of absorption, because there is

[1] See the figures in Strasburger's *Leitungsbahnen*, p. 601.

nothing corresponding to the increased negative pressure
due to continued transpiration in experiment 109.

(112)   *Strasburger's air-pump experiment*[1].

The last experiment depends on a current of water
being drawn through wood by diminished air pressure.
In the following experiment the current moves in opposition to negative pressure.

Cut a sound branch of yew (*Taxus*), peel the lower 8
or 10 cm. and place it in water for about 12 hours; cut a
clean surface and fix it tightly in a perforated rubber
cork fitted into a bottle with a ground mouth.   The cork
is also perforated for a tube connected with an air-pump.
The tube must only just project below the cork, while the
yew branch must be thrust through far enough to dip
well below the eosin solution, which should not more than
half fill the bottle.   When the air-pump is set in action,
it is obvious that its tendency is to suck out the contents
of the tracheids at the cut end, and as a fact air-bubbles
are seen to issue at that point.   Nevertheless, in spite of
this the eosin rises in the branch.   Leave the pump running
for 6 or 7 hours, when the branch should be sawn off
above the cork, without stopping the pump, so as to avoid
injection of the wood with eosin.   Readings of the barometer should be taken in the course of the experiment
and compared with the readings of the pump-manometer[2].

---

[1] *Leitungsbahnen*, p. 795.   A similar experiment is given by Janse,
Pringsheim's Jahrb. 1887.
[2] We have only used a negative pressure of 50 cm., but Strasburger used
72 cm.

SECTION B. **Loss of Water by Transpiration.**

(**113**) *Loss of weight.*

To get a general idea of the amount of loss due to transpiration it is well to take a series of weighings of a plant growing in a flower-pot. Select a plant[1] with a large leaf-surface, in a small flower-pot, so that it may not be too heavy for the balance[2]. In order to confine the loss by evaporation to the plant, the surface of the earth must be covered with a divided disc of sheet-cork painted over with wax-mixture. The pot is wrapped in sheet india-rubber, which may be held in its place as shown in fig. 17; the glass vessel *c* grips the rubber sheet, and also serves to prevent evaporation from the bottom of the pot. Since it may be necessary to water the plant during the course of the experiment, a corked tube must be fitted into a hole in the cork plate.

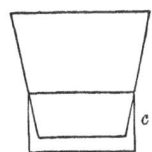

FIG. 17. Exp. 113.

(**114**) *Transpiration compared with evaporation of a surface of water.*

To estimate the transpiration from a given leaf-surface it will be necessary to take a plant small enough to be placed on a more delicate balance. Detmer recom-

---

[1] *Helianthus tuberosus* or *Chrysanthemum.*

[2] We use a French druggist's balance capable of carrying 4 or 5 kilograms, and of turning with 0·5 gram when loaded with 1000 grams in each pan; it is a useful form of balance for the purpose in question. The beam, etc. being below in the box, and the pans therefore free to take a tall plant.

mends a *Phaseolus* grown in a glass vessel having a ground
edge so that it can be covered by a divided glass plate.

We have found it a simple plan to make use of Lambert
and Butler's $\frac{1}{4}$ lb. tobacco tins. A small plant such as a
Pelargonium can be knocked out of its pot and trans-
planted to one of these tins. Owing to the stopper-like
arrangement by which the tins are closed, it is easy to
replace the tin lid by a split cork through which the stem
and a watering-tube pass.

Ascertain the loss by transpiration in say 12 hours,
and at the same time ascertain the loss of weight from
a shallow dish of water of known area.

Now calculate the transpiring area of the plant and
compare its loss of weight per unit of area with that of
the water.

If a planimeter is not available the area may be
calculated by tracing the form of a leaf on stout paper,
cutting it out and comparing its weight with that of a
piece of known area. If the plant has leaves of various
sizes, the leaves are classified into two or three sizes, and
the area of one of each heap is taken. If the amount
of stem or branches is considerable an estimate should
be made of the area of these, and the amount added to
the area of the leaves.

(**115**) *Loss of weight compared with absorption.*

To demonstrate that the loss of weight is roughly
equal to the amount of water absorbed by a cut branch,
select a branch of Portugal laurel (*Prunus lusitanica*)
which has no young growing shoots, or remove any that

may be present.   Let it remain in water for **24** hours, cut
a fresh surface and fit it in a bottle arranged as in fig. 18.
The branch *B* fits a tube which pierces the cork and

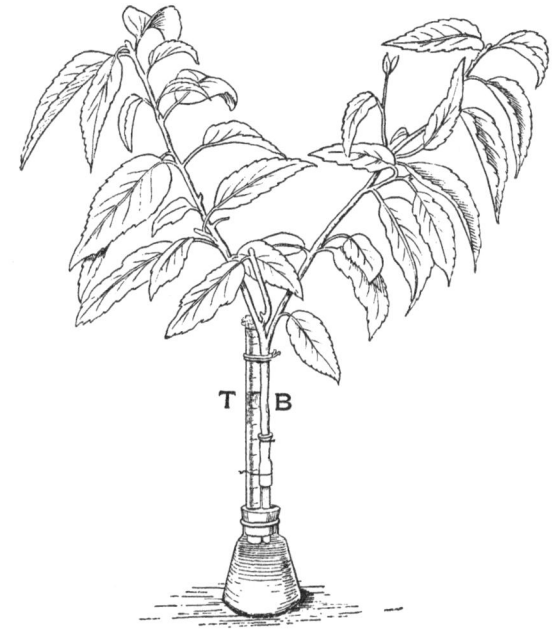

Fɪɢ. 18.   Exp. 115.

should dip well into the water in the bottle[1].   Through
another hole in the cork a tube *T* graduated into $\frac{1}{10}$ c.c.
and holding about 20 c.c. is passed: this serves to record
the amount of water absorbed by *B*: the opening of *T*
must be closed with a plug of cotton-wool to allow air to
enter, and yet to check evaporation from *T*.   In one experiment we found that a branch of laurel with **30** leaves

[1] In fig. 18 the tubes do not dip sufficiently deep.

weighed, with the bottle of water, 287 grams : by using
one of Becker's balances without a glass case, and having
a beam supported 40 cm. above the table of the balance, it
was possible to place the bottle on the pan and arrange
the leaves and branches so as to be clear of the scale-pan
knife-edges.

Weighings and readings should be taken at hourly
intervals. The burette ought to be read to 0·01 c.c., if the
weight is recorded to 0·01 gram. It is instructive to
repeat the experiment with a freshly cut branch in which
the negative pressure is not satisfied. It will be seen
that the absorption is much greater than the loss by
weight, it may, for instance, be three times as great.
After taking a few hourly readings the apparatus should
be left to itself for 12 hours when equality between gain
and loss should be fairly established.

(**116**) *Spring Balance.*

We have found the following arrangement, fig. 19, useful
for demonstrating the loss of weight due to transpiration,
and it is probable that it may prove to be useful for
research purposes under certain conditions. A cut branch
in a test-tube of water T is suspended by a wire to a
spiral spring S. At P the wire passes through a small
hole in a metal plate : at I a fine spun glass filament is
fastened horizontally to the wire. As T loses weight the
index rises and its movement is recorded by means of
a horizontal microscope. With the low power of our
microscope one degree of the ocular micrometer equals
0·044 mm.: the following readings (expressed in gradua-

tions of the micrometer) were obtained by adding a

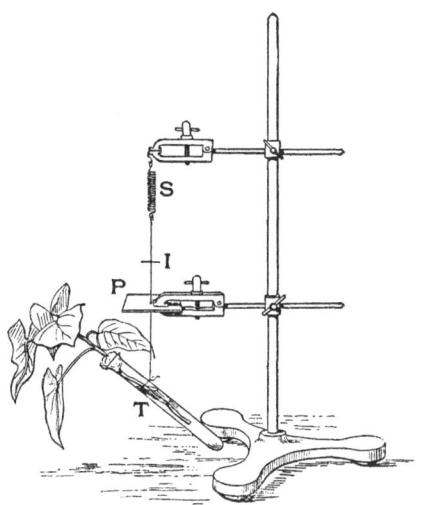

<figure>FIG. 19. Exp. 116.</figure>

decigram at a time to a scale-pan suspended to the
spring,

7·5°, 7·5°, 7·3°, 7·2°, 7·0°, 7·5°, 7·2°, 7·3°, 7·5°, average 7·3°.
When the whole weight was put on at once the index
moved through 66° of the micrometer, giving 7·3° as the
calculated value of 0·1 gram[1].

When a transpiring plant is suspended the loss of
weight may be read every 5 minutes with less disturbance
to the plant and with less labour to the observer than
with a balance.

---

[1] The springs we use are made by Salter of Birmingham : they are
about 5 cm. in length, when unstretched.

SECTION C. **Stomata. Bloom. Lenticels.**

(**117**)   *Stomatal transpiration*[1].

Cut a pair of similar well-grown leaves of *Ficus elastica*, and when the bleeding of latex from the cut ends has practically ceased slip about an inch of tightly fitting rubber tubing over the leaf stalk, leaving $\frac{1}{2}$ inch of tube projecting; then fold the free end down and wire it tightly to the tube-covered stalk. In this way evaporation from the cut end of the stalk is prevented: the wire ties will also serve to hang up the leaves. Having weighed them, hang them up close together in a dry room for 2 or 3 hours, when they must be again weighed. These weighings give the ratio between the normal transpiration of the two leaves. Now smear the lower surface of A and the upper surface of B with vaseline, which should be carefully rubbed on with a finger. Weigh the specimens and leave them for 24 hours. It will be found that B loses in weight something like 10 times as much as A. In one experiment made in winter the leaf whose stomata were closed remained green and fairly fresh for a fortnight, while those with an ungreased lower surface were brown and withered.

(**118**)   *Stomatal transpiration* (*observed by another method*).

For demonstration purposes the well-known experiment of Garreau[2] can be repeated in a very simple manner, with a rough sort of hygrometer represented in the sectional diagram, fig. 18. It consists of a small glass

[1] Garreau, *Ann. Sc. Nat.* 1850.
[2] *Ann. Sc. Nat.* 1850.

cylinder g across the mouth of which a glass tube is fixed:
from the centre of the tube a piece of
Stipa-awn s projects at right angles
and bears at its end an index i,
which may conveniently be made of
thin iron wire. The awn is sensitive
to hygrometric change, in damp air it

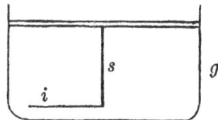

FIG. 20.   Exp. 118.

untwists, in dry air it twists up again.   If the vessel is therefore placed mouth downwards on damp blotting-paper or on
a transpiring leaf, the index i will rotate and its movement
can be read off on a graduated ring of paper fastened to
the bottom of the vessel, or a graduated strip of paper
attached round the vessel close to the bottom.   If two
hygrometers are made, one may be placed on each surface
of a leaf and the difference in the movement of their
indices compared.   Certain precautions are necessary: in
the first place, it is difficult to get two pieces of awn[1]
which behave similarly, so that it is necessary to graduate
the two hygrometers to make their reading comparable.
Take a filter-paper and damp it carefully, making sure
that it is not wetter in one part than the other, place it
on a flat glass plate and having marked the position of
the index with pencil on the paper rings in both hygrometers, place them side by side on the wet paper.   After
from 4 to 8 minutes mark the position of the index again.
If, for instance, the movement of hygrometer A is only $\frac{2}{3}$
of that of B, it is clear that the paper ring on A must be
marked out in divisions each of which is $\frac{2}{3}$ of the unit

---

[1] The awn should be thoroughly ripe, brown in colour, not yellow,
and stiff not weedy in texture.

used for hygrometer *B*. Our hygrometer scales are
usually divided into 60—100 divisions.

To fit the hygrometers on to the leaf (we use laurel
leaves[1]) two plates of cork are wanted, each having a circular
opening slightly .smaller than the hygrometer: one plate
has a groove running across the middle which receives the
midrib of the leaf, and allows it to lie flat between two
plates. One hygrometer is placed mouth upwards on the
table, then the leaf between the plates, then the other
hygrometer mouth downwards: the whole being kept
steady by a weight of 2 or 3 oz. placed on the top. For
laurel leaves 7 or 8 minutes is generally long enough to
wait before reading the hygrometers.

[The general behaviour of stomata may be conveniently
studied here.]

(**118** A)   *Stahl's cobalt method*[2].

It is well known that paper impregnated with a
solution of a cobalt salt, e.g. cobalt chloride, changes from
blue to red when it is placed in damp air and reassumes
the blue colour when dried. Stahl has been able by
taking advantage of this fact to demonstrate a number
of points in the physiology of the stomata. The sensitiveness of the cobalt paper depends on the strength of
the solution employed, for delicate reactions Stahl re-

___

[1] Ivy (*Hedera*) leaves are equally good, and if the apical part of the
leaf is used, the cork plates may be dispensed with. *Sparmannia* leaves
give a good result. In some cases it will be found convenient to attach
the vessels to the surface of the leaf by melted wax-mixture instead of
using the arrangement above described.

[2] Stahl, *Botan. Zeitung*, 1894, p. 117.

commends 1 p.c.; for the following we employ a 5 p.c. solution.

Strips of cobalt paper are placed on each side of a hypostomatal leaf[1] and are covered by glass plates which should project beyond the edges of the paper. Stahl uses in some cases sheets of talc as being less heavy and therefore more easily fixed in place than glass plates. The glass or talc plates being gently clamped at the edges the papers are confined in spaces in which the dryness of the air will depend on the transpiration of the two surfaces of the leaf. The paper on the lower surface reddens rapidly, while that on the upper side remains blue.

A simple plan is to take a pair of similar leaves, placing one, A, with the stomata upwards, the other, B, in the reverse position on a dry folded cloth; after covering them with a strip of cobalt paper, place a sheet of glass over them which makes a good contact with the yielding cloth. After observing the rapid reddening of the paper over A, the experiment should be repeated, reversing the leaves, so that B has now its stomata upwards.

To demonstrate the very small amount of transpiration from the stomatal surface of some leaves we slightly modify Stahl's method by cementing to the leaf surface the glass plates covering the cobalt papers. We use small glass plates about 2·5 cm. × 3·5 cm. made by cutting ordinary microscopic slides in half and attach them by running in melted wax-mixture at the point of junction

[1] That is with the stomata only on the lower surface; Stahl recommends *Tradescantia zebrina*, *Pyrus communis*, *Populus nigra*, &c.

of glass and leaf. Ivy leaves treated in this way give striking results in winter, the paper on the upper side remaining blue for as long as 24 hours.

(**118** B) *Closure of stomata in half-withered leaves.*

For this experiment Stahl[1] recommends *Tropæolum majus* and *Chelidonium majus,* and we have found *Impatiens noli-me-tangere* useful. One out of a pair of similar leaves of *Impatiens* is gathered and allowed to lie on the table ; as soon as it shows obvious flaccidity the fresh leaf is gathered and the two are placed lower side upmost on a soft dry folded cloth, covered with dry cobalt paper and then with a glass plate about 10 cm. × 10 cm. The cloth yields slightly and allows close contact between the surfaces of cloth, leaf, paper and glass without danger of injuring the leaf. The paper over the withered leaf remains blue after the fresh leaf has reddened the other piece of paper.

(**118** C) *Closure of stomata by treatment with salt solution.*

Stahl[2] has shown that young plants of *Acer pseudoplatanus,* seedlings of *Tropæolum majus, Phaseolus multiflorus* and *Zea mais* grown in pots and watered with $\frac{1}{2}$ p.c. NaCl solution exhibit closure of stomata after a few days' culture. Leaves taken from the experimental plants are subject to the cobalt test with similar leaves from plants which have not been watered with salt solution. He describes the stomatal surface of the experimental leaves as not reddening the paper for a long time, whereas the control leaves behaved normally.

[1] *loc. cit.* p. 120.   [2] *loc. cit.* p. 134.

We have only made the experiment in winter, when we found that the leaves on branches of *Prunus lusitanica* which had been in 0·5 p.c. NaCl for 8 days, had distinctly less power of reddening cobalt paper placed on their lower surfaces than had the leaves of the control branches.

(**119**) *Stomata: connection with intercellular spaces*[1].

For this experiment the choice of a suitable leaf is important. The following answer well : *Arum maculatum, Ranunculus ficaria, Eranthis hiemalis, Caltha palustris, Primula sinensis, Limnanthemum sp.*

The leaf stalk is fixed air-tight in a rubber cork which fits a bottle filled with water. The leaf can be fixed by piercing the cork with a hole too small for the stalk, and dividing the cork longitudinally down one side till the hole is laid open. Or an undivided hole may be used if made air-tight with wax-mixture. Through a second hole in the cork passes a glass tube (it must not dip into the water) connected with the air-pump. When the pump is set in action a stream of bubbles emerges from the cut end of the stalk, which is below the surface of the water. The reverse experiment may also be made by placing the lamina in the water and the cut end in the air.

(**120**) *Injection with water.*

Many leaves, e.g. *Ranunculus ficaria, Eranthis hiemalis, Arum maculatum, Caltha palustris, Limnanthemum, Hydrocharis*, can be injected by sucking the stalk

[1] See the figures in Pfeffer's *Physiologie*, Vol. I. p. 96.

with the mouth while the lamina is in water. Or the air-pump may be applied. It is easy to ascertain the pressure necessary for injection by the following arrangement.

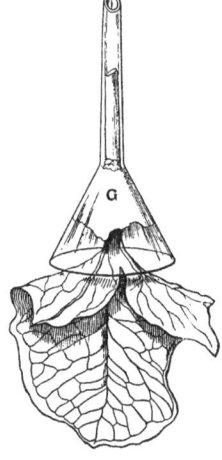

The leaf is attached to one arm of a T tube : by means of the second arm suction is applied, and the third ends in a bent tube dipping into mercury. If the junction between the stalk and the T tube is not quite air-tight it is of no consequence, since the leak affects the leaf and the manometer equally. A good air-tight junction may however be made by Devaux's method[1] of melting the leaf stalk into a funnel with gelatine G as shown in fig. 21.

FIG. 21.   Exp. 120.

(**121**)   *Frost effects.*

The injection of the intercellular spaces with water can be observed on the frozen leaves of certain evergreens. In a hard frost the leaves of the ivy (*Hedera*) have a semi-transparent, dark green appearance, like, but not so dark as, the colour of a water-logged leaf. If the leaf is pinched between the finger and thumb the normal light green colour returns to the under-surface : the same effect may be produced by dipping a corner of the leaf in lukewarm water. If however the whole leaf including the cut stalk is thawed under water, it does not become light

[1] *Ann. Sc. Nat.* 1889.

green, but assumes the very dark tint of an injected leaf. In the first case thawing produces injection with *air*, in the second with *water*. The explanation given by Moll[1] is that the cells of the mesophyll, in freezing, give up water to the intercellular spaces, and that when they are thawed the cells absorb the melted ice from the intercellular spaces, which then fill up with air or water as the case may be.

(122) *Blocking of stomata by water*[2].

One arm of a bent glass tube is gently pushed into the cavity of an onion (*Allium cepa*) leaf and is there firmly secured by a ligature of soft cotton or worsted. The other end of the tube is held in the mouth, and the leaf is immersed in water: by blowing gently, bubbles are forced out of the stomata on the external surface. Now clean the bloom from a zone of the leaf, which may be done by gently rubbing it with a plug of cotton-wool dipped in warm water.

On again immersing the leaf and blowing, it will be seen that the air does not come out of the cleaned zone, which is now thoroughly wetted owing to the removal of the bloom.

When onions are not available the flower stalk of *Narcissus* answers well. A rubber tube can be slipped over the cut end, and the stalk plunged upside down, flower and all, into a jar of water. The stalk should not be cut too near the ground but where it begins to be hollow.

[1] *Archives Néerlandaises*, Vol. xv.
[2] Sachs' *Physiologie*, French Trans., p. 280.

(123)   *Opening and closing of stomata*[1].

The majority of stomata close when surface-sections of the leaf are placed in water. Some leaves however behave in the reverse way: of these the most easily accessible are those which form the floating rosettes of *Callitriche*. If the tissue of the lower surface of the leaf is gently scraped away with a needle, and the leaf is mounted in water with the upper surface upmost the stomata are visible; they can be made to close by irrigation with 2·5 p.c. NaCl solution, and again to open by replacing the salt solution with water.

(124)   *Electric effect*[2].

Strips from the under-surface of the leaf of *Ranunculus ficaria* are mounted *dry* under a cover-glass, on a slide bearing a pair of microscopic electrodes. On passing the induced current the stomata close. A current, slightly stronger than that bearable on the tongue, is necessary. If *Callitriche* is used it must be mounted in water: a stronger current is needed.

(125)   *Lenticels*[3].

The fact that lenticels communicate with intercellular spaces may be conveniently studied in connection with the parallel results obtained with stomata. Fit a woody dicotyledonous branch (dog-wood, *Cornus sanguinea*, does well) to the short arm of a **U** tube by means of firmly wired

---

[1] Mohl, *Botan. Zeitung*, 1856 ; N. J. C. Müller, *Pringsheim's Jahrb.* VIII. p. 75 ; Leitgeb, *Mittheilungen Bot. Institut Graz*, 1886.

[2] N. J. C. Müller, *Pringsheim's Jahrb.* VIII.

[3] Stahl, *Botan. Zeitung*, 1873, p. 613 ; the author recommends *Gingko biloba*, *Sambucus nigra* and *Lonicera tatarica*.

india-rubber tube. The vessels and intercellular spaces at
the upper (free) end of the stick are to be secured by an
india-rubber tube wired on and closed by being folded
down parallel to the branch and again wired. The
U tube is placed in a jar of water so that the stick is
immersed, and mercury is poured into the long arm :
after a varying time air is seen to issue in fine streams
from the lenticels,—that is if they are open. Fifteen or
20 cm. of mercury give sufficient pressure.

(126) *Bloom.*

The character of the leaf-surface has an effect on
transpiration, as may be shown in the following way[1].
Bring into the laboratory a pot of *Kleinia* or *Cotyledon,*
or some other succulent plant with a good bloom. It
is best to bring the pot into the laboratory because
if the leaves are cut before they are wanted they are
likely to get rubbed. Cut off two similar leaves (or two
similar twigs) and pierce each specimen with a piece of thin
copper wire to serve as a hook by which to hang it up.
Paint the cut surfaces with lard: weigh the specimens,
hang them in a dry room for 12 or 24 hours, and weigh
them again. This will give the normal transpiration of
the two specimens. Now remove the bloom from one
specimen by delicate sponging with water at about 35° C.
When it is dry weigh both again, and once more expose
them to the dry air of the laboratory for 12 or 24 hours.
The cleaned specimen should now transpire relatively
more than the other.

[1] See Garreau, *Ann. Sc. Nat.* S. 3, T. XIII., p. 339.

CHAPTER V.

SECTION A. *Imbibition, Hygroscopic movement, Polariscope, Osmosis.*

SECTION B. *Turgor.*

SECTION C. *Tensions of tissues.*

Section A.

(127) *Laminaria,*[1] *microscopic observation.*

The thallus of *Laminaria* is useful for the demonstration of some of the simpler phenomena of imbibition, the increase in size which takes place when the dry tissue is placed in water being very great.

Cut transverse sections of the dry stalk of *Laminaria,* mount them in alcohol, and irrigate them with water, while under the microscope, by placing a drop of water on one side of the cover-glass and a strip of filter-paper on the other. Observe that the cell-walls increase enormously in thickness.

---

[1] *Laminaria* has been much used for experiments on imbibition, e.g. by Sachs in his *Arbeiten,* II. p. 305, and by Reinke in Hanstein's *Botan. Abhand.* IV. 1879.

(128) *Increase of size not uniform in direction.*

Cut a rectangular piece out of the thallus of Laminaria, choosing a part free from wrinkles; let it be slightly oblong so that the longitudinal axis of the thallus may be distinguishable.   Measure the length and breadth with a millimeter scale and mark, by means of a pin-hole in the corner, the two edges along which the measurements were taken.   Place it in water and measure it again in a quarter of an hour.   It will be found to have increased far more in the transverse than in the longitudinal direction.

(129) *Effect of temperature*[1].

Weigh, to 0·1 gram, about 30 grams of air-dried peas : place them in water at about 26° C., and let them remain at that temperature for 2 hours.   Dry them first with a soft cloth, then with filter-paper, and weigh them again. Place at the same time a similar weight of peas in water at 10°—14° C. and compare the gain in weight in the two cases.   The peas, which have been in warm water, will have absorbed from two to two and a-half times as much water as the second lot[2].

(130) *Salt solution.*

Weigh about 30 grams of peas, taking care to use the same material as that employed in experiment 129 ; place them in 10 per cent. NaCl solution, which must be kept at the same temperature as the cool water in experiment

---

[1] See Reinke in Hanstein's *Botan. Abhand.* IV.

[2] According to Nobbe (*Handbuch der Samenkunde*, 1876, p. 230) the effect of temperature is not very apparent when peas are soaked for longer periods.

129. After 2 hours dry and weigh them. The peas will be found to have increased in weight, but much less than the control material in experiment 129.

(131) *Stipa pennata.*

The awn of *Stipa pennata* is, as previously explained[1], extremely hygroscopic, untwisting when wetted, and twisting again when dried. To observe its movements it

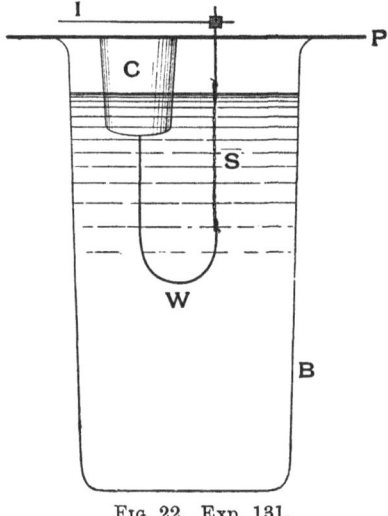

FIG. 22. Exp. 131.

must be fitted up as shown in fig. 22. The awn *S* is lashed with fine wire to a straight wire above, and below to a hooked wire *W*: the latter, *W*, is fixed into the cork *C*, which is attached underneath the stout paper or cardboard *P*: in the middle of *P* is a hole through which the straight wire passes, bearing at right angles the index

[1] Exp. 118, p. 102.

I. On the upper surface of *P* a graduated circle is marked. When the *Stipa*-awn is dipped in water the index moves in the direction of the hands of a watch, which we may call the "wet" direction; when it is removed it will, after a time, reverse itself and move in the "dry" direction.

(132) *Stipa: effects of temperature.*

As in the case of experiment 129, so here, it may be shown that warmth increases the action of water very greatly. Prepare 2 beakers of water, one at 14°—15° C., the other at 40°—45° C.; place the awn in the cold water, and when the index has clearly begun its slow movement, plunge it in the warm water, when the untwisting is at once accelerated.

(133) *Stipa: effects of temperature.*

Place the awn in water at about 15° C. and allow it to come to rest: then transfer it to water at about 40° C., there will be a sudden deflection in the "wet" direction and a return to a position slightly on the "dry" side of the original position of rest.

A similar result may be obtained with a dry awn, by holding it high above a spirit-lamp or a small gas-flame, taking care not to scorch it; the first sudden move will be in the "wet" direction, the heat will then dry the awn and a steady "dry" movement will follow. These effects of temperature are not understood[1].

[1] Francis Darwin, *Transactions of Linnean Society*, 1876. Is it possible that they have something in common with the contraction produced in violin strings by heat? See Engelmann *Ursprung der Muskelkraft*, 1893.

(**134**) *Stipa: salt-solution.*

An awn which has come to rest in water can be made to twist in the dry direction by transferring it to 10 per cent. NaCl solution.

(**135**) *Stipa: mechanism of the movement.*

The twisting power of the awn depends on the hygroscopic torsion of its individual cells. To show this it is necessary to isolate some of the elements. Prepare Schulze's macerating fluid by dissolving in 50 c.c. of nitric acid, 1 grm. of potassium chlorate; add to this half its volume of water, and boil a ripe awn cut into two or three pieces in a test-tube half full of the diluted liquid. It is best not to boil it too much ; as soon as the awn is clearly beginning to disintegrate it must be removed, thoroughly washed in water[1], and teased out with needles. A small portion[2] is now dried on a glass slip without a cover-glass over a flame and examined under the microscope with a low power. Cells will be found which are obviously twisted on their axes, and which at once untwist when water is added. Or the dried fragments of awn can be seen to exhibit torsion if an assistant breathes cautiously on the preparation whilst under observation.

(**136**) *Nobbe's experiment*[3].

Take two stoppered bottles of about 400 c.c. capacity : fit each with a rubber cork through which passes a narrow

---

[1] Because the fumes of Schulze's fluid are bad for the lenses of microscopes.

[2] The colourless pieces should be selected.

[3] *Handbuch der Samenkunde*, 1876, p. 126.

graduated tube. Half fill bottle A with whole peas, and place the same quantity of split peas in B. Fill both bottles with water which has acquired the temperature of the room, and take care to get rid, by shaking, of any air adhering to the peas; force in the corks firmly, and note the height of the water column in each bottle. If the peas increase in volume by the amount of the water absorbed, the level of the water will not change. This is what happens[1] in B, which contains the split peas, but in A the level rapidly rises and then falls. This curious phenomenon is said to be due to the expansion of the testas of the peas producing a temporary increase in size.

(137) *Variability in the swelling of seeds.*

In the case of certain plants, there is great variability in the time required for the absorption of water by the seeds. This is especially the case with leguminous seeds. Nobbe[2] describes the phenomenon in clover seeds; we use those of *Lupinus hirsutus*, which have a rough surface to the testa.

Take 100 lupin seeds and place them in a flat-bottomed vessel (so that they may be easily examined) and add about a liter of tap water. After 24 hours the majority of the seeds will be swollen, the minority which have not yet swollen can be easily distinguished by their smaller size.

The swollen seeds should be removed and the water

[1] Reinke shows that, in some cases of imbibition, the increase in volume is slightly less than the water absorbed.   Hanstein's *Botan. Abhand.* IV. 1879, p. 60.

[2] *Handbuch der Samenkunde*, 1876, p. 112.

renewed to minimise decomposition, and this should be
done at intervals of 12 hours until all the seeds are
swollen, noting at each examination the number of freshly
imbibed seeds.

The cause of the individuality in imbibition seems to
depend, not on the cotyledons, but on the seed-coats.
This may be demonstrated on a similar number of seeds,
after the first interval of 24 hours has elapsed. Pick out
all the seeds which have not swollen, and in half of them
pierce the testa with a needle, leaving the other half
intact. It will be found that all the punctured seeds
swell within 12 hours, whereas only a percentage of the
intact seeds are swollen[1].

(**138**) *Rise of temperature accompanying imbibition*[2].

Prepare enough dry powdered starch[3] to make a layer
2—3 cm. thick at the bottom of a beaker, and place a
similar quantity of water in a second beaker. When
starch and water are at the same temperature, pour the
water into the first beaker, stir with a thermometer bulb,
and note the rise of a few degrees which takes place.

(**139**) *Work done during imbibition.*

Saw out a square inch from a deal board of about
¾ of an inch in thickness. Put it in a flat photographic
dish and let it serve as a support for a 28 lb. weight. On
adding water the wood swells and raises the weight, a

---

[1] See Detmer, *Praktikum*, p. 131.

[2] Nägeli, *Theorie d. Gährung*, 1879, p. 133.

[3] The starch should be dried at 100° C. and may be allowed to cool,
without special precautions, to the room-temperature, when it will still
be sufficiently dry for our purpose.

movement which may be recorded in various ways, e.g. by a horizontal microscope, or by the micrometer-screw described below, experiment 155. The weight will compress the dry wood slightly, so that it is necessary to wait until the index comes to rest before the water is added. The experiment is a rough one, the unsteadiness of the weight on the small bit of wood being a source of error, another being introduced by the unequal swelling of the wood tipping the weight slightly on one side. For this reason we use the following arrangement. The ordinary 28 lb. weights are pierced by a hole so that the surface of the wooden block is visible from above. It is therefore possible to use as an index a vertical wire or a glass filament, the lower end of which rests in a shallow depression in the surface of the wood, while it is kept vertical by passing through a hole in a fixed metal plate or horizontal loop of wire. The movement of the upper end of the glass filament is then observed with a horizontal microscope.

(140) *Observations with the polariscope.*

The apparatus consists of two parts, the polariser and the analyser. In the Zeiss pattern of instrument, the former of these is to be fixed axially on the substage of the microscope above the mirror; the analyser separates into two pieces, one, a disc—which should be graduated—fastens on to the upper end of the tube like a collar; into this an ordinary ocular is slipped, and the other piece is fitted on to it like a cap.

The essential part of each is a "Nicol's prism"—

pieces of doubly refractive Iceland spar so cut and disposed that the light transmitted is all polarised in one particular axial plane. The central upper part of the analyser rotates on the collar-like disc and bears a pointer which records the amount of the rotation. If, with the parts placed in position and the eye at the ocular, the analyser be rotated until its plane of polarisation becomes identical with that of the polariser, the light will be transmitted through it undiminished and the field of the microscope appear bright. On now slowly turning the analyser either way through a right angle, the light will gradually fade until the field is completely dark. In this latter position the polarising planes of the two prisms are at right angles and the analyser intercepts all the light transmitted by the polariser. On continuing the rotation of the analyser through a further 90°, maximum brightness from coincidence of the planes will be again obtained. With the dark field, place on the stage of the microscope a slide on which is mounted some doubly refractive object such as almost any vegetable tissue, and it will be seen that parts of each of its elements will appear black and parts white. The object thus seen is said to be viewed " with crossed Nicols " and the appearance is due to the optical structure of the object so altering the polarisation of the light that part of it and part only is capable of passing through the analyser in its present position. Hence the alternating light and dark markings. On shifting the analyser through a right angle these markings are now seen reversed in their optical properties.

The arrangement of the markings is constant with given objects. Starch grains exhibit on a black field a large black and white Maltese cross with its centre at the hilum of the grain. To see this well, starch grains of potato should be mounted in Canada balsam so as to be transparent and when examined in ordinary light quite invisible. Advantage may be taken of this method for detecting small doubly refractive bodies, otherwise difficult to perceive. Thus the very minute crystals of calcium oxalate occurring in many leaves are often difficult to make out, but if the leaf be decolorised and rendered quite transparent by soaking it in strong chloral-hydrate solution, the distribution of these crystals may be easily observed by their light crossed markings on a black field. This method was employed by Schimper[1] in his interesting researches on the physiology of calcium oxalate. An example of its use has been given in experiment 78 (p. 66) where the amount of oxalate occurring in leaves under various conditions is studied.

(141) *Tension.*

Crystalline bodies are anisotropic, and it is one view of the significance of the anisotropism of organised bodies, cell-walls, starch grains, etc., to attribute this to a crystalline structure of the ultimate particles—micellæ of Nägeli—of which these bodies are held to be built up.

Another view denies this crystalline structure and attributes the anisotropism to the tensions or strains

[1] Schimper, *Bot. Zeitung*, 1888, p. 81.

existing in the substance of the cell-walls or the starch grains as a result of their particular structure. In order to realise that tension may produce double refraction in a substance that is not in itself anisotropic, for example glass, a fine glass filament should be forcibly extended, while it is under observation in polarised light in the field of the microscope. To make the effect more apparent use should be made of the selenite discs generally supplied with the polarising apparatus. The various colours which bodies exhibit when viewed in these coloured fields give a measure of the strength of their double refraction. To perform the stretching experiment a piece of glass rod, drawn out at the blow-pipe to a fine filament in its middle part, should be so clamped at one end that the fine part lies across the field of the microscope and can be focussed with a low power. The selenite disc No. I., which gives a red-purple field, should be suitably placed below the glass thread, which then appears as a double black contour with red-purple between. The free end of the glass rod should rest on some guiding support which will keep it in focus but allow it to be stretched. If the observer looks down the microscope while the rod is steadily pulled, the colour of the centre of the thread will be seen to change distinctly, the nature of the change depending on the amount of traction exerted upon the filament: on releasing it the purple colour re-appears.

Compression should give a different colour-change but this cannot be easily exhibited. Other transparent bodies such as gelatine films show similar effects.

(142) *Traube's artificial cell*[1].

Traube's method is of great interest as a graphic way of demonstrating the possibility of pressure arising osmotically inside a cell. The method is moreover capable of giving results of great value, especially as modified by Pfeffer[2]. The following experiment is merely meant to serve as a demonstration.

Fill a beaker with a solution (2 or 3 per cent.) of potassium ferrocyanide and drop it into a fragment of copper chloride or acetate[3]. The copper salt is instantly coated with a precipitated membrane of copper ferro-cyanide.

In the artificial cell so produced osmotic pressure arises by which the brittle cell-wall is broken, but is instantly mended by the formation of a fresh precipitate : as soon as the wall is mended the pressure inside again increases, and again ruptures the cell-wall, and thus by a series of breaks, healed as soon as made, an apparently continuous growth of the cell takes place.

(143) *Slowness of diffusion*[4].

Fill a tall narrow jar with water and with the help of

[1] Traube in *Archiv für Anatomie und Physiologie* (Reichert and Du Bois-Reymond), 1867. [2] *Osmotische Untersuchungen*, 1877.

[3] In the first edition of this book I recommended copper sulphide for Traube's experiment,—a piece of advice which somewhat puzzled my friends and reviewers. We have been in the habit of using for experiment 142 a sample of commercial copper sulphide because the cells which it forms "grow" better than those produced by $CuSO_4$. And I re-commended its use without ascertaining that (as turns out to be the case) the result is entirely due to the presence of a soluble copper salt in our sulphide. [F. D. 1895.]

[4] See de Vries, *Bot. Zeitung*, 1885, p. 1.

a long funnel run in very slowly and carefully a stratum of concentrated solution of potassium bichromate, which accumulates at the bottom of the jar.

It will be seen that the colour spreads to the upper strata with extraordinary slowness. The chief physiological interest of the result is that it serves to suggest the value, to the living cell, of protoplasmic circulation.

(**144**)   *Relation of membrane to diffusing fluid*[1].

A dialyser made of vegetable parchment is filled with a 1 per cent. solution of di-sodic phosphate coloured with methylene blue, and is placed in distilled water; after some hours the blue colour is visible in the water. If, however, a precipitation membrane of calcium phosphate is produced in the wall of the dialyser, the methylene blue is unable to pass. The precipitate is produced by immersing, in 1 per cent. calcium nitrate, a dialyser filled as before with 1 per cent. di-sodic phosphate coloured with methylene blue. The importance of the experiment is to show that by the formation of a precipitation membrane the osmotic quality of the parchment is changed.

(**145**)   *Absorption of methylene blue.*

It is interesting to note in connection with the last experiment that methylene blue, as Pfeffer[2] has shown, can pass a living protoplasmic membrane.

Two or three sprigs of *Elodea* are placed in about a liter of tap water containing 0·0008 per cent. of methylene

---

[1] Taken from Detmer's *Praktikum*, p. 96.
[2] *Untersuchungen aus dem Bot. Institut zu Tübingen*, II. p. 223.

blue, after from 24 to 36 hours the living cells will be found to contain blue cell sap.

SECTION B. **Turgor.**

(146) *Plasmolysis, microscopic observation*[1].

In order to realise the existence of turgor the well-known microscopic observation of the effect of salt solution on turgescent tissues should be repeated. Plasmolysis is easily seen in *Spirogyra*, or any tissue with coloured cell sap may be used; it is only necessary to irrigate a preparation with 5 % NaCl solution. It is instructive to compare the result of plasmolysis with the change produced by death. In the first case the cell sap remains within the protoplasmic sac, in the killed cell it escapes and moreover stains the dead protoplasm.

(147) *Recovery after plasmolysis.*

It is important to realise that plasmolysed parts are in no way injured, and that they recover their normal condition when the plasmolysing fluid is replaced by water. A few simple observations on roots of *V. faba* serve for this purpose. A bean root 2—3 cm. in length is placed in 5 % NaCl solution, where it almost immediately becomes soft and flaccid. When replaced in water it quickly becomes turgid again[2].

---

[1] De Vries, *Untersuchungen über Zellstreckung*, 1877.

[2] We have observed that the root of the bean, if placed alternately in salt solution and water several times, becomes translucent, being in fact injected with water. It would seem that the collapse and returgescence of the cells act like a pump and fill the intercellular spaces.

The observations here suggested are meant as illustrations of the very simplest aspect of turgor, chiefly to show that turgor is an osmotic phenomenon, since the condition of the cell is clearly regulated by the relation between the cell sap and the environing fluid.

(148) *Osmotic strength of cell sap in terms of* KNO₃.

The method of de Vries[1] depends on the fact explained in experiment 163 (Section C) that when a turgescent shoot is bisected longitudinally each half curves outwards, i.e. with the epidermis on the concave side. If the curved portions are put in water the curvature increases greatly: if they are placed in strong NaCl solution (5%) they uncurl, i.e. become straight again, or they may even become convex on the epidermic side. Therefore an intermediate strength of salt solution must be discoverable which equals the cell sap in osmotic force, and which neither produces increase nor decrease in curvature.

In summer we use the scape of the dandelion, *Taraxacum*; in winter the hypocotyl of *Ricinus* seedlings.

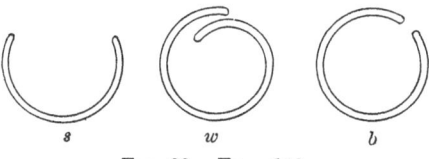

s w b

FIG. 23. Exp. 148.

The dandelion is split longitudinally into four strips which, on being dipped for a moment into water, curl up into

[1] *Pringsheim's Jahrbücher*, XIV.

spirals and can then be cut up into some 7 or 8 rings, *b*, fig. 23 : these are delicate tests of changes in turgescence since a small increase or decrease in the curvature of the turgescent tissue is at once perceptible.   Thus *s* is in too strong a solution, *w* is in too weak a solution, while *b* is in one that almost exactly balances the osmotic power of the cell sap.

The process with *Ricinus* is a little more troublesome ; the hypocotyl is split in 4 or more longitudinal portions, and the form of each is traced with a paint-brush (which answers better than a pencil) on paper.   We now have a number of curved bits of tissue (whose form is known)

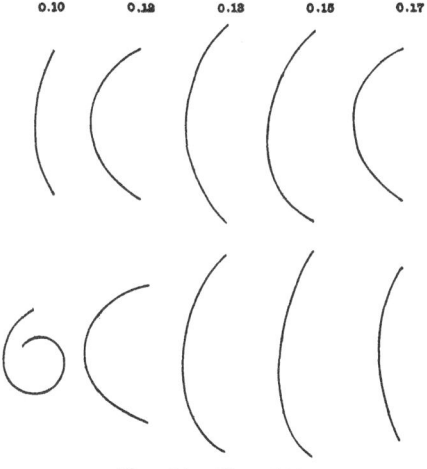

FIG. 24.   Exp. 148.

each one of which must be placed in a solution of a different strength.   These solutions are made according to equivalents, and in the case of KNO_3 (which forms the standard) may contain 0·05, 0·10, 0·11, 0·12, 0·13, 0·14 gram-molecules per liter; stronger solutions may however be needed. After a quarter of an hour the result may be noted: if the material consist of dandelion rings the result is obvious on inspection; with *Ricinus* the segments must be compared with the sketches.

Fig. 24 gives tracings of pieces of split *Ricinus* hypocotyl before and after immersion.

The upper row of tracings gives the form of the pieces before being placed in the solutions, the lower row shows the change of form produced by the immersion. The numbers 0·10, 0·12, etc. give the strength of the KNO_3 solution in which each was placed.

It will be seen that the first two have increased in curvature, while the last two have uncurled and the middle piece (i.e. that in the 0·13 solution) remains unchanged. Therefore 0·13 expresses the osmotic quality of the cell sap in terms of KNO_3.

(149) *Isotonic coefficient.*

The same experiment must be made with cane-sugar, using solutions 0·16, 0·18, 0·20, 0·22, 0·24. From the results obtained (combined with those of experiment 148) it is possible to calculate the isotonic coefficient (I. C.) of cane-sugar, i.e. the attraction for water of a molecule of cane-sugar expressed in terms of the attraction of a molecule of KNO_3 for water. For the sake of convenience the value of this last quantity is taken as 3 instead of 1. We have then the following calculation. Assuming that we have found that the cell sap$=0.13$ KNO_3 and also$=0.20$ cane-sugar,

$$\frac{\text{I. C. of sugar}}{3} = \frac{1.3}{2.0} = 0.65,$$

$$\therefore \quad \text{I. C. of sugar} = 1.95,$$

or in round numbers $= 2$.

In this way, using the plant as an index, it is possible to ascertain the osmotic intensity of solutions of a number of substances in relation to a living protoplasmic membrane.

(**150**) *Microscopic method.*

The principle of de Vries' second method is simple: small portions of tissue are put in a graduated series of salt solutions and the equivalence between one of them and the cell sap is estimated by the degree of plasmolysis observed microscopically. The tissue must contain coloured cell sap so that plasmolysis may be readily observed. De Vries recommends as material the epidermis of parts of the leaf of *Tradescantia discolor, Begonia manicata,* or *Curcuma rubricaulis.* Of these *Tradescantia discolor* is the most universally available and is the only one of which we have any experience. In *Tradescantia* the part of the leaf used is the epidermis of the under-surface: to get good results it is necessary to use closely adjacent parts of the epidermis taken from the midrib. De Vries makes parallel incisions $1\frac{1}{2}$ or 2 mm. apart in the epidermis of the midrib: the areas so marked out can then be freed by a surface-cut with a razor. The fragments of the epidermis must remain in the solutions for at least an hour before being examined. The condition of each is noted as P. completely plasmolysed, H. P. half plasmolysed, or N. P. not plasmolysed.

The solution which produces the H. P. effect is taken as osmotically equivalent to the cell sap.

(**151**) *Estimation of the hydrostatic pressure in turgescent tissue*[1].

Take an actively growing flower stalk such as that of the cowslip (*Primula veris*), which must be in the budding condition: mark off 100 mm. near the upper end and place the stalk in 5 % NaCl solution. As soon as it is thoroughly flaccid it should be measured again, when it will be found to be shorter, owing to the elastic contraction of the cell-walls, which were previously stretched by the turgescence of the cells. If it can now be ascertained what force is needed to stretch the shrunken stalk to its original length, we shall know what was the force exerted by the turgidity of the tissues.

The bud of the cowslip is fixed in a screw-clamp lined with cork-plates and the clamp is fixed to a horizontal board, so that the stalk will be stretched when the other end is pulled. The basal end of the stalk may be simply knotted to a piece of cord, which passes over a pulley let into the board, and supports a scale-pan. A millimeter scale having been arranged so that the distance between the marks on the stalk can be easily read off, weights are added to the scale-pan until the marks are once more

[1] De Vries, *Untersuchungen über die mechanischen Ursachen der Zellstreckung* (1877), p. 118.

100 mm. apart. The diameter of the stalk must be roughly measured, and the area calculated, so that the force which is equivalent to the hydrostatic pressure in the tissues may be expressed in grams per square millimeter. It should finally be expressed in terms of atmospheric pressure,—which equals about 10 grams per sq. mm. Something between 3 and 6 atmospheres may be expected as the result.

(**152**)   *Pfeffer's gypsum method*[1].

Pfeffer has devised a method of estimating the pressure exerted by growing plants of which we have no practical experience: the following description is taken from his paper.

The principle will be understood from fig. 25. The cotyledons and the basal part of the radicle are contained in the pot *n* and kept damp by means of sawdust. The extremity of the root is contained in the two blocks of gypsum *a* and *b*, so that as the root grows *a* and *b* are separated. Since *a* is fixed against the pot *n*, the block *b* moves, and in doing so compresses the oval spring *f*. The degree of compression, and therefore the force exerted, is estimated by reading, with a horizontal microscope, the distance between the needle-points fitted to the inside of the spring.

The following is the method of fitting the plant into the apparatus.

The seedling bean is placed in the flower-pot *n* filled with damp sawdust so that 15—30 mm. of the root

[1] Druck- und Arbeitsleistung, &c. *Abhandl. d. k. Sächs. Ges.* Bd. xx. 1893.

project through the hole. A lid is placed on the pot, which
is turned upside down and the root (which projects

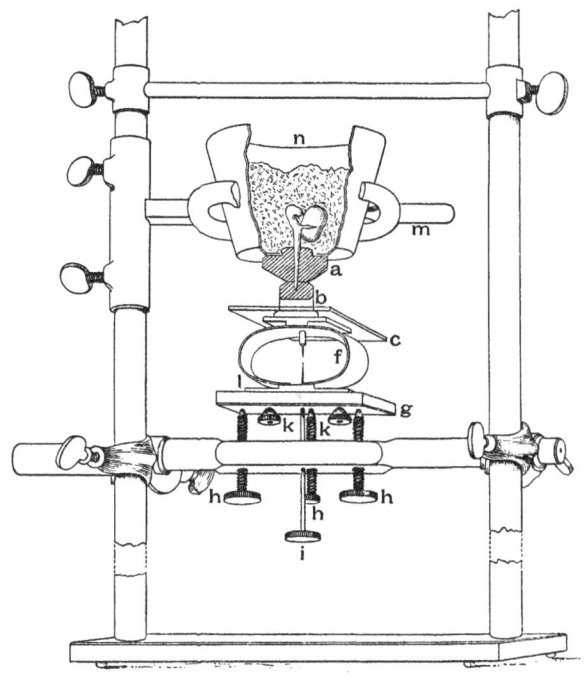

Fig. 25. Exp. 152. Copied from Pfeffer.

vertically upwards) is covered with soft gypsum. A piece
of waxed paper in which a hole has been made (with a
hot needle) is slipped over the tip of the root and pressed
down with a bored glass plate. In this way the block of
gypsum a is formed; when it is sufficiently set, the waxed
paper is removed, and for it is substituted a piece of wet
tissue-paper on which the block b is added. The form of the

blocks a and b is regulated by cylinders of paper acting as moulds. When block b is set hard it may be removed from the root, trimmed with a knife and replaced: at the same time the tissue-paper may be removed. Before the flower-pot is placed in the supporting ring m the block of gypsum b must be secured in its place by tying it with a thread which will be cut when the arrangement is complete. The block b is fixed by fluid gypsum to the glass plate c which rests on the spring f.

The plate l, forming part of the spring, is fixed by the small screws k, k to the solid plate g, which can be raised and lowered by means of the screws h, h, h. In this way the desired amount of pressure can be applied at the beginning of the experiment. The distance between the needle-points is regulated by the screw i which moves the lower needle.

SECTION C. **Tensions of tissues.**

(153) *Longitudinal tensions.*

The fundamental experiment illustrating the condition of strain or tension[1] which consists in turgescent tissues may be made in summer or spring on any rapidly growing juicy shoot, e.g. elder (*Sambucus*), or with certain leaf-stalks, e.g. that of the rhubarb (*Rheum*). In winter it is sometimes difficult to find suitable material: if a green-house is available, the leaf stalks of *Richardia* will answer well. It is best to get fairly long shoots, i.e. not less than

[1] See Sachs' *Text-book*, Sect. 14, 15. The whole discussion should be studied.

20 cm., so that measurements to 1 mm. may give perceptible results. The material must be as fresh as possible, and if it has to be brought from any considerable distance must be wrapped in a wet cloth and placed in a vasculum : in this case too, it is worth while to take care that the vasculum is held vertically, lest the shoots should take a geotropic curvature, as they may do if kept horizontal for an hour.

Place the shoot on the table, cut the ends as square as possible and measure its length with a millimeter scale placed lengthwise on it. Remove a strip of cortical tissue along the side measured; it will be shorter than the original shoot. Now remove the whole of the cortical tissue, and measure the length of the cylinder of pith remaining, which will be found to be longer than the intact shoot.

This experiment shows that the internal tissues are in a state of compression, while the cortex is extended. It is important to note that the amount of extension of the freed pith need not by any means be the same as the contraction of the cortex[1]. If the experiment is repeated with a scape of *Fritillaria imperialis* which has ceased to grow, it will be found that the pith lengthens considerably while the contraction of the cortex is very slight.

(**154**)   *Extension of pith in water.*

When pith is placed in water it increases greatly in length in consequence of the increased turgescence of its

---

[1] Sachs' *Text-book*, p. 797.

cells[1]. To show this, place the pith from experiment 153 in water, and measure it again after an hour.

(155) *Change in the transverse dimensions of pith*[2].

Cut from the fresh turgescent pith of the stem of the *Helianthus, Sambucus* and of *Impatiens sultani,* also from a rhubarb leaf stalk, parallel-sided pieces about 10—15 mm. in length and 5 mm. in width, taking especial care that they are free from all cortical tissue. Place a piece on its side (i.e. with the 5 mm. dimension vertical) in a small flat-bottomed glass vessel, and lay on the pith an ebonite vessel measuring 4—5 mm. in diameter by 2—3 in depth, and containing oil. By means of the following arrangement the oil is made to serve as a delicate index of any shrinking or swelling of the pith. A vertical micrometer screw graduated to 0·01 mm. carries at its lower end a vertical needle, which can be lowered until it dimples the polished surface of the oil; the moment of contact is sharply defined, and in this way changes of 0·01 mm. in the diameter of the pith are easily read. After taking a few readings, which usually indicate a slight shrinking, water should be added. The results of the increased turgescence so produced vary with the material employed; in the case of *Sambucus* and *Helianthus* the pith begins to shrink, i.e. diminish in transverse diameter; Rhubarb-pith increases and afterwards diminishes; while *Impatiens* increases but does not diminish.

[1] See A. Bateson and F. Darwin " On the Effect of Stimulation on Turgescent Vegetable Tissues," *Linnean Society's Journal*, xxiv. 1889.

[2] See Miss Anna Bateson, *Annals of Botany*, Vol. iv. p. 117. A drawing of the micrometer screw is given in Chapter vi. fig. 27.

(**156**) *Change in tangential dimension.*

Cut with a dry rasor sections (such as would be considered very thick for microscopic purposes) of the fresh scape of the dandelion (*Taraxacum*) or (in winter) of the hollow hypocotyl of *Ricinus.* Place the rings, so prepared, on a glass plate, and with a scalpel divide each at one point. The divided rings are now placed in water, when their curvature is found to increase, the curling inwards being due to the shrinking in tangential direction of the turgescent tissue forming the inner part of the ring.

(**157**) *De Vries' experiment on the shortening of roots.*

In the roots of certain plants a phenomenon has been observed by de Vries[1] which seems to be of the same character as those described under experiments 155—56. The roots shorten along their longitudinal axes when turgescence is increased, and lengthen when turgescence is diminished, e.g. by immersion in 5 per cent. NaCl solution.

De Vries describes the phenomenon in *Carum, Dipsacus* and other plants.

Full directions are given by Detmer[2] for observation on the roots of young (2—3 months) plants of *Carum carvi.* If suitable material is wanting for following out Detmer's instructions, it is generally possible to find roots in which shortening has already occurred, and which are remarkable for their wrinkled exterior. The roots of hyacinths grown in water show the phenomenon well.

---

[1] *Landw. Jahrb.* ix. 1880.
[2] *Praktikum*, p. 248.

(158) *Imperfect elasticity of plant-tissues.*

The fact that the tissues of a growing shoot or leaf are extensible, but not perfectly elastic, can be demonstrated on a variety of material, e.g. a flower scape of *Polyanthus,* or the leaf of a *Narcissus*: the form of the last-named makes it convenient for the purpose. For this and similar experiments a strong sheet of cork mounted on a board is convenient: one end of the leaf is clamped between the mounted cork and a free block of cork, in such a position that the other end of the leaf projects beyond the board. Two marks about 100 mm. apart are painted on the leaf, one being close to the clamped end. The distance between the marks having been read on a mm. scale (clamped to the cork-board) the projecting end of the leaf is pulled with the hand ; the distance between the marks is now to be read off without diminishing the traction, and again when the leaf is left to itself. The leaf will be found to be permanently extended ; the temporary and permanent extensions should be recorded in percentages of the original length.

(159) *Cyclometer*[1].

Take a straight turgescent shoot, e.g. a young cabbage-shoot, bend it forcibly, and then release it : it will be found to have taken on a permanent curve. This is only another way of demonstrating what is shown in experiment 158 : the cortical tissues on the convex side of the shoot are forcibly elongated by the bending, and being imperfectly

[1] Sachs' *Text-book*, p. 784.

elastic do not return to their original length, thus producing a distortion of the shoot.

To get an accurate notion of what occurs in this
experiment it is desirable to measure the radius of
(1) the curvature forcibly produced, (2) of the permanent
curvature remaining. This may be done with Sachs'
cyclometer, which consists of a number of concentric
semicircles drawn on a board. By applying the shoot to
the board, and comparing its outline with the semicircles,
the radius of curvature of the shoot can be approximately
ascertained and noted. In our laboratory we have two
boards, one bearing semicircles of which radii range from
1 to 20 cm. in length: while the radii of the arcs on the
other board range from 21 to 45 cm.

(160)   *Hofmeister's experiment*[1].

This is in principle the same as experiment 159; it
has, however, a certain classic interest which makes it
worth repeating.

What is needed is a vertical turgescent shoot fixed
firmly at its lower end : it may be either a plant growing
in a pot, or a shoot fixed into a clamp by its basal end.
In either case the base of the shoot is smartly struck with
a light stick so as to produce violent curvature of the free
end of the shoot towards the side which is struck. The
consequence is the same as that in experiment 159,
namely, that a permanent curvature is produced in
consequence of the overstretching of the convex side of
the shoot.

[1] *Berichte d. k. Sächs. Gesell. d. Wiss.* 1859.

(161) *Loss of rigidity.*

The rigidity of a turgescent shoot is dependent on (among other factors) the resistance of the cortical tissues; if, by overstretching, these are permanently lengthened, the rigidity of the system is lessened.

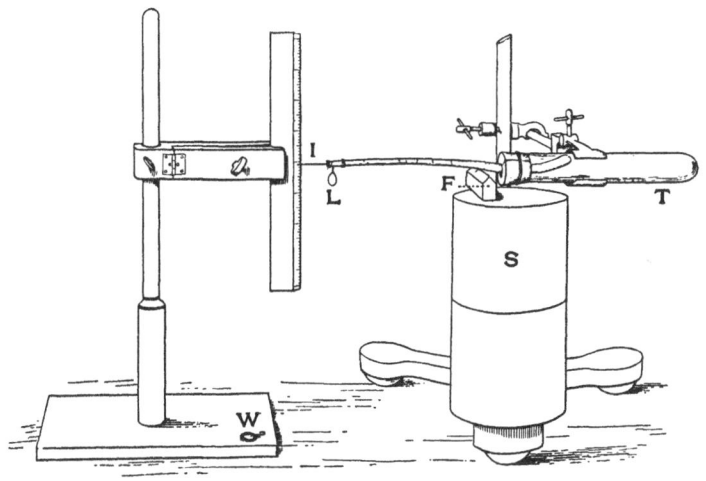

FIG. 26. Exp. 161.

A straight turgescent shoot is fixed firmly by means of a bored and split cork in a test-tube of water, T, figure 26, and at a point, which should be marked by a streak of Indian ink, it is further supported on a prism of wood, F, resting on a support S. At the free end, the shoot bears a needle acting as an index I, and a loop of wire L, to which weights may be hung. Having noted the position of I on the scale, hang a small weight W (a coil of

lead wire of 8—10 grams) on the loop, and read off the
position of the index. Remove the weight, and bend the
shoot two or three times backwards and forwards in the
vertical plane. When the weight is once more attached,
the index will move through a greater distance than that
at first recorded.

(**162**) *Increased length.*

Since the pith is in a state of compression, any increased length of the cortical tissue must result in an
increase in the length of the whole shoot. Therefore
bending a turgescent shoot backwards and forwards as in
experiment 161, must lengthen it. The length must be
accurately measured, say to 0·1 mm., to make sure of a
result.

(**163**) *Splitting turgescent tissues.*

The relation between the compressed pith and the
stretched cortex can be demonstrated by dividing a shoot
longitudinally. It is best to prepare the shoot by cutting
it flat on two opposite sides, making a slab of pith
bounded on two sides by strips of cortical tissue : this is
placed on a glass plate and bisected with a knife, when
each half curves so that the pith is on the convex, the
cortex on the concave side. The curvature can be greatly
increased by putting the half-shoots in water. This
increase is strikingly seen if a scape of dandelion (*Taraxacum*) is split into 4 or 5 longitudinal strips, which curl
up in water into spirals of many turns[1].

---

[1] This fact has already been utilised in experiment 148, p. 126.

(164) *Splitting a root.*

In some turgescent structures the erectile (compressed) tissue is external, while the resisting or stretched tissue is internal. In such cases the result of splitting longitudinally must obviously be the opposite of that just described: the parts will curve inwards, towards the longitudinal axis, not away from it.

Pull up a seedling bean (*V. faba*) with a root 3 or 4 cm. long, split the apical centimeter with a scalpel, and put it in lukewarm (25°—30° C.) water. The halves will certainly not curve outwards, and will after a little time show a slight inward bend. The aërial roots of Aroids show the same tensions.

(165) *Splitting a pulvinus.*

Take a large pulvinus of *Phaseolus* and cut from it an axial slab as described under experiment 163. Split the slab down the central strand, and put the halves in water, when they will curve inwards, i.e. with the vascular tissue on the concave side.

CHAPTER VI.

SECTION A.   *Conditions   of   growth   (experiments   without   special   apparatus).*

SECTION B.   *Distribution of growth.*

SECTION C.   *Auxanometers.*

SECTION A.   **Conditions.**

(166)   *Method.*

Many experiments may conveniently be made on the radicles of leguminous seedlings.   We use principally the bean (*Vicia faba*), the pea (*Pisum sativum*), as well as *Phaseolus multiflorus*.   All these seeds should be placed in water for 24 hours at least, or until by the tenseness of the testa they show themselves to be thoroughly soaked. They should then, as Sachs recommends[1], be washed with fresh water and placed to germinate in moist sawdust, cocoa-fibre, or powdered peat.   The washing may conveniently be done by placing the seeds in a colander

[1] *Arbeiten*, I. p. 386.

under a running tap for a minute. The sawdust or other
material should be prepared for the reception of the seeds
in a flat vessel of galvanised iron, 50 cm. in diameter and
6 cm. deep, in which the dry sawdust is placed and is
gradually moistened, mixing it thoroughly with the hands
as the water is added. Large flower-pots will serve for
germination of the seeds; they should be loosely filled
with damp sawdust, and when the seeds have been added
they may be covered with crockery plates or sheets of
glass. Beans seem to thrive best at a temperature of
about 15° C., peas and *Phaseolus* may have a slightly higher
temperature.

Beans should be placed in the sawdust with the plane
of the cotyledons vertical and the hilum downwards: the
radicle thus grows out without curving, and the seeds are
ready for cultivation as shown in fig. 33, Chap. VII., where
they are supported on pins stuck into the cork lining of the
cover of a jar. Peas must be pinned through both cotyledons, it is therefore necessary to let them germinate
with the plane of the cotyledons horizontal; the same
position is convenient for *Phaseolus*.

(**167**)   *Free oxygen necessary.*

Place 6 peas in water: when they are soaked remove
the testas, measure the length of the radicle in each, and
pass the peas up into an inverted test-tube of mercury,
as described in exp. 6, p. 10, on intramolecular respiration.
After from 12 to 18 hrs. measure the radicles, which will
be found either not to have grown perceptibly or to a very
slight degree.

(**168**)   *Respiration necessary*[1].

Pick out 12 germinating beans with roots 20–25 mm. in length.   Having gently dried the roots by stroking them with the torn, feathered edge of a piece of filter-paper, mark each at 10 mm. from the tip, by painting a transverse line with good Indian ink[2].   When the ink is dry, place the seeds in water for a few minutes to allow the roots to recover from the effects of the dry air of the room.   Or the total length of the radicle may be measured from a pin-hole, blackened with ink, in the triangular flap of the testa.   Then impale the seeds on " blanket pins," fixing 6 in one jar (A), 6 in another (B).   Fill A with water so that the cotyledons are completely covered, while B is only half filled and the roots allowed to grow in damp air.   After 24 hours remeasure.   The growth of the roots in A will be much smaller than (e.g. one-fifth of) that in B.   It is a remarkable fact that the roots of the bean are not able to obtain enough air from the water, but are dependent on their cotyledons.   For another experiment on this point see experiment 185.

(**169**)   *Effect of salt solution*[3].

Proceed as in exp. 168, but let the water in jar A be sufficient to cover the roots and just to touch the hilums

----

[1] Sachs, in his *Arbeiten*, I. p. 408.

[2] Black photographic varnish may also be used, but the marks often become loose from water getting under the crust of varnish.   On the other hand Indian ink becomes faint in colour in water.   We have with advantage followed Pfeffer (*Druck- und Arbeitsleistung* &c. 1893, p. 294) in the use of Bormann's " unauslöschbare schwarze Tusche."

[3] See Stange, *Bot. Zeitung*, 1892, p. 342.

of the seeds: in jar B place a similar amount of 1 p.c.
NaCl solution.　Measure again after 12 or 18 hours, when
the retarding effect of the salt solution should be obvious.
The roots in B only grow about half as much as those
in A.

(**170**)　*Growth at various temperatures.*

A good rough notion of the effect of temperature may
be obtained by using the seeds of beans or peas.　If a
considerable number of seeds are germinated, it is easy
to find 40 peas whose radicles, just emerging from the
micropyle, are of fairly uniform length.　They are to
be sown, in 4 lots of 10 each, in small flower-pots.　The
pots being covered with glass plates or saucers, are placed
(under otherwise uniform conditions) at temperatures of
39°—40° C., 35° C., 23° C., and at some fairly low temperature, such as 10°—12° C.　The first three temperatures
can easily be kept fairly constant by means of thermostats,
the lower temperature may present difficulties at certain
times of the year.　We employ a hollow-walled box, through
which a rapid current of tap water is allowed to run.

After 48 hours measure again: the average growth of
the radicles at 10°—12° C., 23° C., 35° C., will probably be
in ascending series, while the growth at 39°—40° C. will be
less than that at 35° C.　In one of our experiments the
average length of the radicles was after 48 hours:—

At 10° C.	5 mm.
21	10	„
31	25	„
39·5	15	„

SECTION B. **Distribution.**

(**171**) *Distribution of growth in roots*[1].

Pick out a germinating bean with a root about 2 cm.
long. Make 5 marks, 2 mm. apart, on the root, the
first mark being 2 mm. from the tip of the root-cap.
To mark the root, the seed should be pinned to a cork
plate and the millimeter scale raised on a layer of cork
pinned to the first named cork plate like a step, so that
the graduated edge of the scale can be brought close
to the surface of the root to be marked. Roots easily
suffer from dry air, it is therefore advisable to place each
seedling in water for a few minutes after it has been
marked, when it may be pinned, with 2 or 3 others
similarly treated, in a jar half filled with water, the roots
being in damp air.

After 12—15 hours measure the distance between the
marks. The result will show that the most rapidly growing part of the root is a short distance behind the tip.

(**172**) *Distribution of growth in air-roots.*

In aërial roots the region of growth is of much greater
extent. Mark the air-root of an Aroid (e.g. *Philodendron*)
at intervals of 5 mm. for a space of 30 mm. from the tip;
white paint, such as Aspinall's Enamel, is useful for
marking the dingy coloured roots of these plants. Measure
again after 2 days.

[1] See Sachs' *Arbeiten*, I. p. 414.

(173) *Distribution of growth in flower-stalks, etc.*

For this purpose the scape of the cowslip (*Primula veris*) is useful: select a straight-growing stalk, of which the flowers are still in bud, gather it carefully (by cutting, not by pulling), and mark it at intervals of 5 mm. Keep it in a corked test-tube, with a little water at the bottom, for 12—18 hours.

Vigorously growing shoots of valerian may be treated in the same way, i.e. cut and grown in damp air.

Plants of *Phaseolus* in pots, having 2 or 3 internodes developed, are also useful: the marked internodes should not be cut, but left on the plant.

Here as in the case of the root we get evidence of the "grand period"; the youngest part of the stems has grown but little, then comes a region where growth is more vigorous, and further back growth again becomes less marked. The maximum of growth will probably be at a region which was originally 50 mm. from the apex.

(174) *Grand period; time observation*[1].

The grand period may be observed with bean roots, 4 or 5 being measured simultaneously. The experiment should be started when the roots are 5 mm. in length, and the length measured from a mark on the cotyledon. A measurement is to be made every day at a given hour.

Or a scale may be fixed parallel and close to the root and the daily increment noted by reading off the position of the tip of the root on the graduations.

[1] Sachs, *Text-book of Botany*, Ed. II. p. 817.

(175) *Growth and plasmolytic shrinking.*

De Vries[1] has shown that in many cases the shrinking produced by plasmolysis is distributed in space in the same way that growth is distributed. In other words, that region of a plant-member which is growing most quickly shrinks most when plasmolysed. But this appears not to be universally the case, as Schwendener and Krabbe[2] have shown. The most striking exception to the parallelism between growth and plasmolytic shrinking is afforded by roots.

Mark a bean root at 5, 10, 15, 20 mm. from the apex and place it in 10 p.c. NaCl solution until completely flaccid. On remeasuring it will be found that plasmolytic shrinking extends considerably further back than (as we know from exp. 171) growth is found to occur.

We find that a convenient method of measuring the distance between marks is the following. The root is laid on wet blotting-paper to prevent it withering and is supported on a table sliding in a horizontal slot, on which a millimeter scale with a vernier is engraved. A small reading microscope with cross wires is fixed vertically above the root, the table is pushed along the slot and the vernier is read as each mark on the root comes under the cross wires. We thus read to 0·1 mm. with fair accuracy.

---

[1] *Zellstreckung*, 1877.
[2] *Pringsheim's Jahrbücher*, xxv. 1893, p. 323. See also Sachs, in his *Arbeiten*, I. p. 396.

SECTION C. **Auxanometers.**

(**176**) *Methods*[1].

Instruments for measuring growth are of two kinds:
(1) those in which the continuous presence of the observer is necessary, and (2) self-recording instruments.

Of the first class various simple forms may be constructed. If a cord attached to the summit of a flower-
stalk is passed over a pulley (supported vertically above
the plant) and attached to a weight, the descent of the
weight in a given time equals the elongation of the
plant. The descent of the weight may be read in various
ways, but in all cases certain sources of error have to be
avoided. If cord is employed, it should be of fine plaited
silk, because twisted cords vary greatly in length with the
moisture of the air. The alteration of lengths of plaited
cord may, as Sachs points out, be to a larger extent
prevented by oiling or waxing. It is however in all ways
better to use fine flexible wire. The weight must only be
sufficient to keep the cord or wire thoroughly tight
because any serious strain interferes with growth. The
cord may be attached by a simple knot or loop, and if
there seems any danger of its cutting the tissues the stem
may be protected by a strip of gummed paper wrapped
round it before the cord is attached. If wire is used, it
should be hooked to a piece of soft cord tied round the
plant. Any shrinking or swelling of the earth in the pot
will obviously introduce errors. These can only be avoided

[1] Sachs, in his *Arbeiten*, I. p. 113, and in his *Text-book*, Ed. II. p. 826.
See also Baranetzky, *Mém. Acad. St Pétersbourg*, XXVII. 1879.

by watering the plant thoroughly at the beginning of the experiment, and leaving it unwatered during the rest of the observation; the shrinking of the earth will thus be spread over a considerable period. The most serious error however is the curvature of the stem, which may be either spontaneous or more frequently due to heliotropism. If the plant can be illuminated from above, so much the better; if not, a large bright mirror must be placed close behind it, which neutralises one-sided illumination. In our experiments we use principally the flower scape of *Narcissus*, which is but slightly affected by lateral light.

(**177**) *The descent of the weight measured on a scale.*

The simplest plan is to fix an index to the weight and read its movement on a vertical scale. A piece of sheet-lead 15 mm. × 20 mm. folded across the middle will serve as a weight, and a fine sewing-needle placed horizontally in the fold of the lead can be secured in its place by hammering the lead. In this way the growth can be estimated in 0·1 mm., and the arrangement might be used for measuring the daily and nightly growth of a plant for a series of days.

(**178**) *Micrometer screw.*

The weight in this case bears a vertical instead of a horizontal needle, and its descent brings the needle into contact with a surface of oil or mercury contained in a cup which is raised or lowered by means of a micrometer screw[1]. The moment of contact is a very definite

---

[1] This method was suggested and the apparatus designed by Mr H. Darwin, of the Cambridge Scientific Instrument Company.

one, and it is not difficult to read to 0·01 mm.   If oil is the
fluid used, the vessel must first be lowered until the
needle-point hangs clear, the screw is then cautiously
raised until the bright surface of the oil is dimpled at the
point of contact.   If mercury is used a vertical wire or

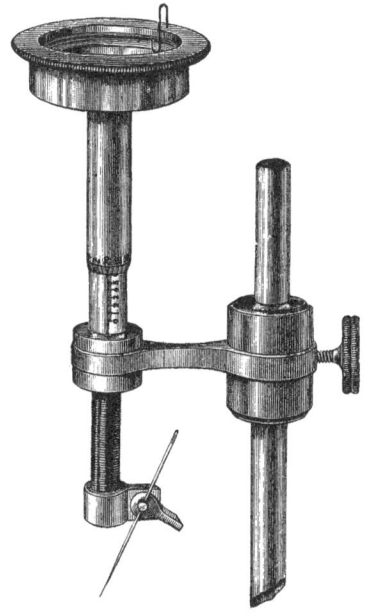

FIG. 27.   Exp. 178.
From the Cambridge Instrument Company's *Catalogue*.

thread must be hung close to the micrometer, between it
and the light, and the moment of contact of the needle-
point and the mercury determined by the distortion of the
image of the thread seen by reflection in the mercury.   In
either case the apparatus must stand on a steady table.

Fig. 27 shows the micrometer; it bears at its lower
extremity a needle which, as explained above (exp. 155,
p. 135), is useful for various measurements. The hook on
the edge of the cup gives a means of knowing when the
oil in the cup is horizontal. If this is the case the point
of the hook will remain in a constant relative position,
with regard to the surface, as the screw is turned. Thus
if the cup is filled so that the point just dimples the surface,
that state of things should be continuous during rotation.

(**179**) *Arc-indicator.*

The cord from the plant passes round an easily
moveable pulley, and ends in a small weight. As the
plant grows the pulley turns, and its rotation, magnified
by means of an index projecting radially from the pulley,
is read off from time to time on a graduated arc. The
instrument is described and figured in Sachs' *Text-book of
Botany*, 2nd edit., Engl. Tr., p. 826.

(**180**) *Microscope.*

The special merit of the microscopic method is that by
its use the attachment of a cord to the plant is rendered
unnecessary. The plant therefore grows in normal conditions, moreover it is now possible to keep the plant in
constant slow rotation about a vertical axis, and thus to
avoid heliotropic curvatures; lastly by this means delicate
plants, e.g. moulds, and delicate plant-members, e.g. roots,
can be observed[1]. We use a horizontal microscope

[1] The microscopic method as designed by Sachs is described and
figured in Vines' paper in Sachs' *Arbeiten*, II. p. 135. The horizontal
microscope may of course be used to read the descent of the weight in
experiments of the type of 177, 178.

designed by Mr H. Darwin of the Cambridge Scientific
Instrument Company, which has a wide field (10 mm.), a
long focal distance (about 23 cm.), and reads to 0·1 mm.;
also one by Albrecht of Tübingen, whose lowest power
gives 0·044 mm. for each division of the eye-piece micrometer, with a focal distance of 6 cm. and a field of 4 mm.
in diameter.   The microscope can be raised and lowered
by a micrometer screw ; and can, by another screw, be
slowly rotated on the vertical axis,—a movement which is
very convenient.

(**181**)   *Self-recording auxanometer.*

This instrument is described and figured by its in-
ventor, Sachs, in his *Arbeiten*[1].   The principle is well
known, namely, that the descent of the weight, magnified
by means of a lever, is recorded on a drum rotating on a
vertical axis.   In Sachs' paper, p. 116, is an interesting
facsimile of the smoked paper on which a plant has
written its hourly growth.

Sachs uses a drum with continuous motion which,
from being excentric about its vertical axis, only comes in
contact with the index at regular intervals of time, for the
rest of the time the index is free from all constraint, and
this is an advantage.   The instrument which we use was
designed by Mr H. Darwin[2].   The drum (shown in fig. 28)

[1] Vol. I. p. 113 ; also *Text-book of Botany*, Engl. Tr., Ed. II. pp.
827–28.

[2] It is on the same general principle as that of Baranetzky (see the
figures in Vines' *Physiology*, p. 399).   It was however invented independently : we have in the Cambridge Laboratory a drum constructed
by Mr H. Darwin about 1876.

measures 25 cm. in height and 512 mm. in circumference :
it rotates on a vertical steel spindle, and by means of a
clockwork escapement makes a short sudden rotation at
intervals of one hour or at shorter intervals if required.
At each rotation the index writes a short horizontal mark

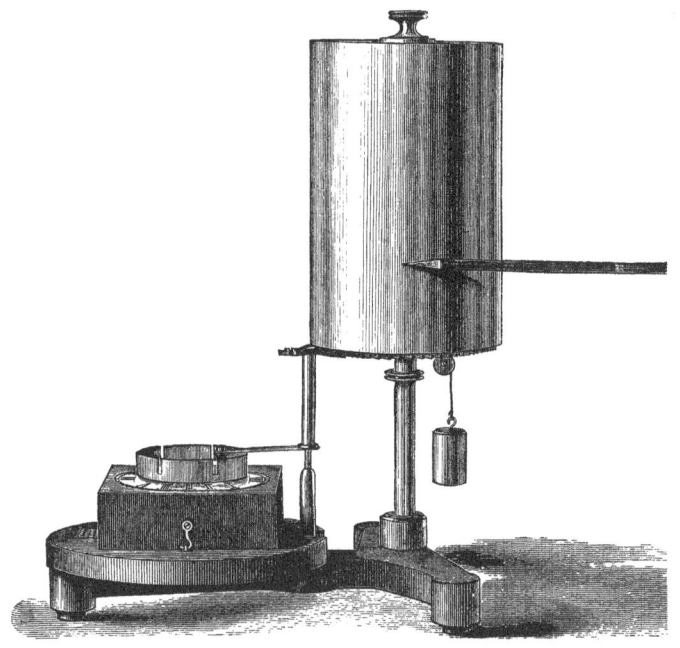

Fig. 28.   Exp. 181.
From the Cambridge Instrument Company's *Catalogue*.

on the smoked paper covering the drum : between whiles
the index moves downwards over the surface of the drum
at a rate proportional to the rate of growth.   The figure
traced by the index is thus a series of steps of which the

"fall," or vertical distance between the horizontals, is proportional to the growth per hour. The slight shock due to the sudden rotation of the drum is an advantage because it prevents the index sticking on the smoked paper. We use a simple lever to magnify the growth of the plant. It measures 605 mm., so that when arranged to multiply by 10, the short arm is 55 mm. in length. The fulcrum is a knife-edge, and the cord connecting the short arm with the plant is also attached by a knife-edge attachment: the index is a piece of platinum foil or a thin piece of quill. Since the vertical displacement of the end of the short arm equals the growth of the plant, the *vertical* distance between the horizontal lines on the tracing must be measured, not the distance along the arc described by the index. It will be seen that an error is introduced by this method: since the short arm describes an arc it is obvious that the cord connecting the lever with the plant does not remain absolutely vertical; but if the cord be of reasonable length, for instance 25 or 30 cm., and the short arm of the lever be also relatively long, e.g. 5 cm., the error is insignificant. For some further details in the use of the auxanometer see exp. 186.

(182)   *Simpler form of recording auxanometer.*

If the lever attached to a growing plant is allowed to write on a smoked surface, and if to this surface a slight shake is given every hour, a record of the hourly growth is obtained. Such an arrangement however only serves for recording growth, or any change which always varies in one direction. If the recording method is used for geo-

tropic after-effect or for the movements of sleeping plants
(as described below) the drum described in exp. 181 must
be used.   The following is a simple method of fitting up a
recording surface.   Take a Bath-Oliver biscuit tin, or any
other sufficiently tall cylindrical box: pierce two holes
just below the point to which the cover overlaps: pass a
stout wire through the holes and support the wire so that
the box may hang with its long axis vertical and its lower
end close to the table.   To the lower end fix a cork with
sealing-wax, and into the cork thrust vertically a strong
short pin.   Take a cheap American clock and fix it to the
table in such a way that the axis on which the hand
moves is vertical, while the hand itself moves in a horizontal
plane.   The clock is so placed that at each hour the hour-
hand just touches the pin fixed to the biscuit box and
causes the box to oscillate on the supporting wire.

(**183**)   *Temperature: microscopic method.*

Select a bean root about 20 mm. in length, impale it
on a pin and fix it to the cover of a flat-sided plate-glass[1]
vessel of water, so that the root is immersed and the hilum
just touches the surface.   Having carefully levelled the
microscope, focus the tip of the root-cap, and note the temperature of the water and the time of the observation.   If
the root-cap is slimy and ragged it may be gently cleaned in
the fingers.   The reading will gain considerably in sharp-

---

[1] A convenient arrangement is to use one of the rectangular glass
vessels used in museums and to cut out by the sand-blast a hole 40 mm.
long by 25 wide on one side, which is afterwards closed by a large
cover-glass fixed inside with Canada balsam. In this way good defi-
nitions can be insured for the microscope.

ness if the root is illuminated horizontally by means of an oblique mirror. Bean roots are so slightly heliotropic[1] that there is no objection to a lateral light. Other roots, e.g. the apheliotropic radicles of *Sinapis*, can only be observed when rotated on a vertical axis. Spontaneous curvatures do however occur in the bean root, and may spoil the experiment : that known as Sachs' curvature which occurs in the plane of the cotyledons is especially troublesome[2]. Askenasy[3] uses maize roots for his growth experiments; he notes that the circumnutation of the roots is in some cases very considerable.

Read the position of the tip of the root after 10 or 15 minutes; and again after a like interval of time ; if the rate of growth is fairly uniform for the two periods, the temperature of the water may be at once raised, if not further readings must be taken. The water may conveniently be siphoned out and replaced by warmer water. It is best to use water 10° or 12° C. warmer than that in the vessel. The growth of the root will be at once accelerated, becoming nearly three times as quick. It will again fall as the water sinks to the temperature of the room.

(**184**)   *Temperature : microscopic method.*

Repeat exp. 183, substituting water at 3° or 4° C. for the water in which the observations were begun. Note the sudden and serious check to growth.

---

[1] According to Kohl (*Die Mechanik der Reizkrümmungen*) bean seedlings grown on the klinostat show distinct apheliotropism.

[2] *Power of Movement in Plants*, p. 91.

[3] *Deutsche botan. Gesellsch.* 1890.

(185) *Respiration : roots.*

Proceed as in exp. 183, and when the growth is fairly uniform fill up the vessel so as to cover the cotyledons. Note the rapid fall in the rate of growth, and the recovery when the cotyledons are again exposed to air.

(186) *Temperature : stems.*

The effect of warmth on growth can be well shown with the recording auxanometer. In fitting up the apparatus the following points must be noted. Let the paper

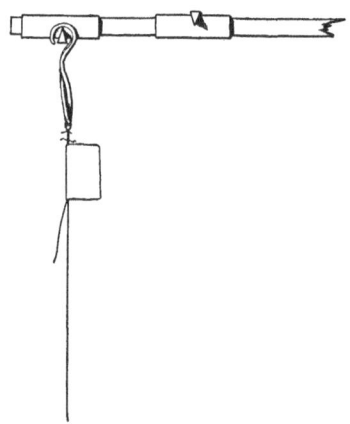

Fig. 29.   Exp. 186.

be smooth and tight on the drum, to insure which the unglazed side of the paper should be wetted with a sponge before it is attached. Do not smoke the paper too heavily : a paraffin stove-lamp, with the long wick turned up so as to flare, may be used for the smoking. The drum should be accurately vertical, which may be tested by a spirit-level placed on the top. The stand which supports the fulcrum of the writing lever must also be vertical. Unless these conditions are fulfilled the index will write at the top of its course, but will leave the surface, or press too hard against it, lower down. The fulcrum should be opposite the middle of the drum, i.e. equally distant from the upper and lower edges of the smoked paper. When the experiment begins the index must touch the paper near its upper edge. To insure this the length of the string joining the short arm of the lever to the upper end of the plant must be regulated. As it is difficult to tie a knot at exactly the desired place, the following plan, fig. 29, should be followed. Let an assistant hold the lever up at the desired angle; pass the string or wire from the plant over the wire loop hanging from the short end of the lever : pull it tight and secure it by a folded piece of paper smeared on the inside with very dense and quickly drying shellac varnish; there will be time to regulate the length before the varnish dries, and when it is dry it is fairly secure. To make it still more secure one or more ligatures of fine silk may be added above the paper, but this is hardly necessary. If after all the index is not at the right height, the fulcrum must be slightly shifted up or down the supporting rod. The cord from the plant to the lever must be vertical, which can be insured by shifting the position of the flower-pot.

Take a *Narcissus* growing in a flower-pot [1], and having a scape in active growth,—the flower being still a young

---

[1] *Narcissus* will however grow quite well from a bulb dug up and kept damp.

bud. Place the pot on a layer of damp sand at the
bottom of a galvanised iron cylinder so supported that a
flame can be placed under it when it is desired to expose
the plant to a higher temperature. The manipulation
necessary in starting the experiment will probably interfere with the growth of the scape, so that it should be
allowed to grow for 3 or 4 hours without further disturbance. A spirit-lamp or very small gas-flame is then
lighted under the cylinder, and the temperature of the air

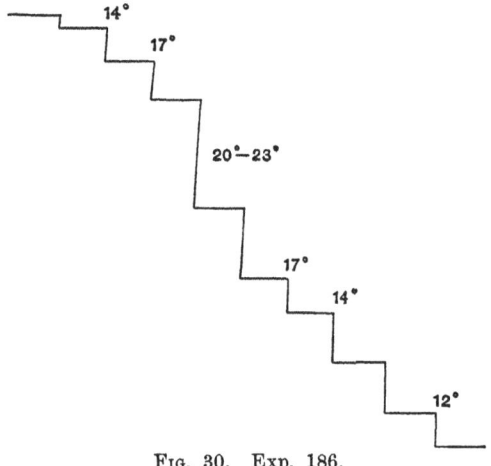

FIG. 30.   Exp. 186.

within the cylinder carefully watched for some little time.
After 2 or 3 hours during which, if the temperature has
been raised by some 10°, the steps traced on the drum
will greatly increase in vertical height,—the flame is
extinguished so that the fall in rate of growth may be
recorded.   Fig. 30 gives a copy of a tracing taken as here

described: the numbers give the temperatures (C.) to which the plant was exposed.

(187) *Light.*

Fit up a *Narcissus* as above described in the dark room, and after 2 or 3 hrs. take down the shutter so that the plant is illuminated; no sunshine must be admitted, otherwise the temperature of the room will be affected. The period of illumination should last 3 hours, when the shutters should be once more closed.   In this experiment it is important to take readings of the wet- and dry-bulb thermometers occasionally, both during the dark and the light period.

(188) *Light.*

Vines[1] has shown that light has a retarding effect on the growth of *Phycomyces.*   A ripe sporangium is allowed to burst in a watch-glass of water and a few drops are placed, by means of a needle, on a thick slice of bread, which should be previously steamed to roughly sterilise it. The bread is placed in a saucer containing a little water and covered with a flat-sided glass cover.   It is now placed on an apparatus by which it is kept revolving once in 30 minutes on a vertical axis, so as to avoid heliotropic curvatures[2].   If the hyphæ are growing vigorously readings may be taken every 15 minutes for an hour, and afterwards at intervals of 30 minutes.   When sufficient readings have been taken to indicate the course of the growth-rate, i.e. to ascertain whether the rate is steady,

[1] Sachs' *Arbeiten*, II. p. 133.

[2] We use a drum turning on a strong steel axis and driven by an endless band connected with a pulley driven by clockwork.

or increasing or decreasing, the culture is darkened by a thick cardboard cover placed over the glass. The rotation should be continued and temperatures taken by a thermometer of which the bulb is inside the glass vessel covering the fungus. It is a good plan to wet the cardboard cover on the outside, so that the temperature during the dark period may be slightly cooler than during the period of illumination. After half-an-hour or an hour the dark cover is removed and the fungus allowed to grow in light for an hour, during which its growth should be noted once or twice. The rate of growth in the dark should differ from that in the light by something like 20 p.c.

(189) *Light.*

The effect of light on the growth of roots should also be demonstrated on the apheliotropic roots of *Sinapis alba*, on account of the interest of the fact in relation to the theories of heliotropic curvature[1]. The seedling mustards are supported by plugs of cotton-wool in holes in a cork plate so that the roots dip in water.

The details of the experiment are practically the same as in exp. 188, but the periods of light and dark should be longer, say 2 or 3 hours.

(190) *Periodicity.*

A *Narcissus* kept for 24 hrs. in the dark room will record its periodic changes in growth-rate. This will probably not be evident on merely inspecting the tracing on the drum; a curve representing growth must therefore be drawn on a fairly large scale.

[1] Francis Darwin, *Ueber das Wachsthum negativ heliotropischer Wurzeln.* Sachs' *Arbeiten,* II. p. 521.

CHAPTER VII.

CURVATURES.

SECTION A. **Geotropism.**

(191) *Region of curvature coincident with region of growth.*

If the curvature is to be observed in damp air a bean-
root of not more than 15 mm. in length answers best.
It should be marked at intervals of 2 mm. and pinned
to the lid of a stoppered jar half full of water; the jar
should be occasionally inverted for a moment or two so
as to thoroughly wet the seedling and the cork. After
12 hours measure the distance between the marks,
which gives the region of greatest growth, and note the

11—2

position of the greatest curvature. It is however better
to let the root grow in damp sawdust behind glass.
The glass wall is not vertical but slopes at an angle of
about 80°; the advantage of the sloping wall is that the
root, in its attempt to grow vertically down, is closely
pressed against the glass, and is visible from the outside.
The marks must be made on the side which comes against
the glass. The figure p. 387 of Sachs' paper (*Arbeiten*, I.)
should be consulted.

(**192**) *Region of curvature.*

In summer any rapidly growing vertical shoots will
serve. Cabbage-shoots answer well, also the stems of
valerian[1]. They should be marked at intervals of 10
mm. and may be either fixed by means of bored and split
corks in bottles of water, or into an embankment of wet
sand in the angle of a tin box, the air being kept
thoroughly damp by wet sand sprinkled over the whole of
the bottom. At a temperature of about 20° C. curvature
will be well marked in 3 hours, when the form of curve
should be noted and the distance between the marks
measured. A strip of thin sheet-lead makes a useful
scale by which to measure the distance between the
marks, since it can be easily applied to the curved surface,
and retains its curvature. Rothert[2] recommends strips
cut from the paper used for Richard's Thermograph. The
lines are said to be fine and at accurately equal distances.

---

[1] In winter young sunflower seedlings may be used.
[2] *Ueber Heliotropismus.* Breslau, 1894, p. 29.

(**193**)   *Subsequent change in the form of curvature*[1].

If specimens prepared as in exp. 192 are left undisturbed for some hours longer, the free end of the shoot will be carried far beyond the vertical. A record of this movement and the subsequent return to the vertical may be made, by using a box with one vertical glass wall; if the shoot is fixed (in a sand-embankment) so that it is close to the glass, its changes of form can be traced with a paint-brush filled with black varnish on the glass. Sachs' large diagrams published with Vol. III. of his *Arbeiten* should be consulted.

(**194**)   *Grass-haulms*[2].

Cut a grass-haulm which is vertical and in which the pulvinus has not curved. Mark the pulvinus on two opposite faces by means of dots some 2 or 3 mm. apart. Having measured the distance between the marks, push the haulm horizontally into a sand-embankment in a tin box so that one set of marks are above, the other below. After 24 hours or more the haulm will have bent at the pulvinus, when the marks must be again measured. The lower surface will have increased greatly in length while the upper surface has become shorter.

Note the pale colour of the lower half of the pulvinus; and if the pulvinus is a hairy one, note the divergence of the hairs below and their convergence above.

[1] Sachs' *Collected Papers*, II. p. 967 (from *Flora*, 1874).
[2] Sachs' *Collected Papers*, II. p. 958 (from the *Arbeiten*, I.).

(**195**)   *Noll's experiment* [1].

Noll's method is to fix a grass-haulm into a glass tube
so narrow that it only just fits, and to leave it horizontal
in damp air.   The pulvinus cannot curve, but the lower
half of the pulvinus grows out into curious excres-
cences.   We arrange the experiment differently.   Cut
a groove in a block of cork about 5 mm. wide and
5 deep.   Arrange 3 or 4 grass stalks on the cork so that
they lie across the groove with the pulvinus of each over
the groove.   Fix them in place by two sheets of cork, one
of which pins down the stalks on one side of the groove,
while the second sheet fixes the parts of the stalks on the
opposite side.   The whole arrangement is put in a tin
box half-full of wet sawdust for 5 or 6 days, when the
result should be clear.

(**196**)   *Free oxygen necessary.*

The bean, which should have a short root (15 mm.), is
pinned in the vertical position to the lower surface of a
rubber cork fitting tightly into the ground neck of a
small bottle.   Hydrogen is passed through the bottle for
about 2 hours to replace the contained air.   The bottle is
then placed on its side so that the root is horizontal.
The tube connecting with the hydrogen-generator is kept
open while the outflow tube by which the hydrogen
current escaped from the bottle is shut; in consequence
of this arrangement the only leakage that can occur is an
escape of hydrogen.   After 24 hours, during which no

---

[1] Sachs' *Arbeiten*, III. p. 509.

geotropic curvature should occur, the hydrogen is replaced by air, when the root will bend downwards.

(197) *Free oxygen necessary.*

The same result may be more simply obtained by keeping seedling beans completely submerged while control specimens are in damp air, or just touching the surface of the water. The geotropic curvature is absent in the case of the submerged specimens.

(198) *Johnson's experiment*[1].

The interest of this experiment (in which a root does external work during geotropic curvature) is now somewhat historical. Its original object was to demonstrate " the unsatisfactory nature of the theories proposed to account for the descent of the radicle[2]," i.e. to show that the root does not bend by mere plasticity. A method of performing the experiment is shown in Pfeffer's *Physiologie*, Vol. II. p. 320, fig. 36.

A similar experiment may be more simply arranged in which the resistance is given by a spring. A pin is driven vertically into the inside of the lid of a jar, and from the lower end of the pin a thin copper wire projects horizontally; at the end of the wire a microscopic cover-glass is cemented so that it lies horizontally. A bean is now pinned to the lid so that its root projects horizontally and rests on the glass-cover; as the root curves down it overcomes the elasticity of the wire. The cotyledons and

---

[1] *Edinb. New Phil. Journal,* 1829, p. 312.

[2] From the title of Johnson's paper.

base of the root should be kept damp by a strip of filter-paper hanging over the seed like a rider and dipping into the water below.

(199)   *Pinot's experiment*[1].

This experiment is the same in principle as the last, but the resistance is supplied by the descent of the root into mercury. The arrangement of the experiment is shown in Sachs' paper on the growth of roots[2]. A shallow dish of 6—8 cm. in diameter is filled with mercury to a depth of 2—3 cm., on which is a layer of water 5—6 mm. in thickness. A split cork is firmly jammed like a rider on the edge of the dish, and to it a bean is pinned so that the root lies horizontally in the water just touching the surface of the mercury. The whole arrangement is covered with a bell-jar and left for 24—48 hours. Another way of fixing the bean, which we find convenient, is simply to support the pin, on which the cotyledons are impaled, in a clamp attached to a small heavy stand. We usually keep the cotyledons wet with a strip of filter-paper dipping into the water.

(200)   *Knight's experiment*[3].

Sachs figures[4] an apparatus which any one can construct for himself, and by which it may be demonstrated

[1] *Ann. Sci. Nat.*, series 1, T. xvii. 1829 (Bibliography, p. 94). See also Hofmeister in *Pringsheim's Jahrbücher*, Vol. iii. p. 105.

[2] *Arbeiten*, i. p. 452, fig. 14.

[3] *Phil. Trans.* 1806.

[4] *Physiologie* (French Trans.), p. 124.

that the geotropic parts of plants bend in relation to centrifugal force.

Another simple plan is to use a water-wheel driven by a strong fine jet of water directed against the wheel from the water-tap. The wheel should stand in a sink fitted with a cover; in this way,—with the help of the spray from the wheel—the experimental plants are kept thoroughly damp.

We use an apparatus designed by Mr H. Darwin. A disc covered with a thick layer of cork is attached to a horizontal axis turning on bicycle ball-bearings. It turns with ease and is driven at considerable velocity by an endless band from a turbine. The experimental plants are kept damp by a bell-jar which is not attached to the revolving disc, but fits by its broad ground edge against a fixed vertical metal plate, through which the axis passes. The space in which the plants rotate is not therefore absolutely closed, but the air can be kept sufficiently damp for practical purposes. The most serious drawback to the apparatus is that the plants are subjected to a current of air produced by their own rotation. This evil has been fairly well overcome by a four-armed fan attached to the disc, and dividing the space inside the bell into four compartments; as the fan rotates, the air within the bell-jar is carried round with the plants.

To use the apparatus it is only necessary to pin seedling beans so that the root lies tangentially; each bean must be fixed on two pins firmly driven into the cork. They should be fixed near the circumference of the disc, but it must be remembered that the roots will curve away from the centre of rotation; allowance must therefore be made for their growth in that direction. Roots curve perfectly well even when covered by a layer of wet sponge pinned completely over them, an arrangement which insures their being kept damp.

The scape of *Taraxacum* or cabbage-shoots may be used for apogeotropic curvature. Each shoot is fixed in a bored cork through which, and through the contained shoot, two strong pins are forced into the cork.

It is well to devote one quarter of the space inside the bell-jar to a piece of dripping wet sponge pinned firmly to the cork; this serves to keep the air moist.

The apparatus should be made to turn at 600 or 700 revolutions a minute which gives a centrifugal force equal to about six times gravity[1], when the experimental plants are at a convenient distance from the centre.

In from 12 to 24 hours a good result should be obtained.

In Pfeffer's laboratory an instrument is used[2] which has the advantage of giving a high centrifugal force without excessive rapidity of rotation. The rotating body being 150 cm. in diameter, it is possible to fix plants at various distances, up to 75 cm. from the centre of rotation, and thus to experiment with a graduated series of stimuli at the same time.

---

[1] It is convenient to keep a table from which the centrifugal force can at once be calculated from the revolutions per minute and the distance of the plant from the centre of rotation.

[2] See Fr. Schwarz, in *Untersuchungen aus dem botan. Institut zu Tübingen*, I. 1881, p. 57.

(**201**) *Sudden curvature*[1].

When a growing shoot is prevented from curving apogeotropically, the gravitation stimulus nevertheless produces some change, so that when freed from constraint the shoot suddenly bends upwards.

The constraint may be applied by placing the shoots horizontally on a shallow layer of damp sawdust, and keeping them down with a sheet of plate-glass. Or they may be fixed to a sheet of cork by pins crossing over the shoot like an X, one such fastening being placed at each end and one in the middle; the cork must then be placed in damp air for some 6 hours, when the plants may be unpinned.

(**202**) *After-effect.*

A turgescent shoot is fixed by means of a cork into a bottle of water so placed that the shoot projects horizontally. A needle to serve as an index is fixed in the free end of the shoot and its position recorded on a vertical scale. After about an hour,—or when the shoot has begun to curve apogeotropically,—the bottle is rotated on its axis through 180°, so that the plane of curvature remains vertical, but what was the upper side of the shoot is now the lower. The index will now travel downwards over the scale, owing to the continuance of the curvature induced by the gravitation stimulus. It will finally come to rest and will at last curve up in the opposite direction[2].

---

[1] Sachs' *Arbeiten*, I. p. 204.
[2] Sachs' *Collected Papers*, II. p. 966 (from *Flora*, 1874).

(**203**)   *After-effects recorded on a rotating surface.*

The movements described at the end of exp. 202 may be made to record themselves in the following way.

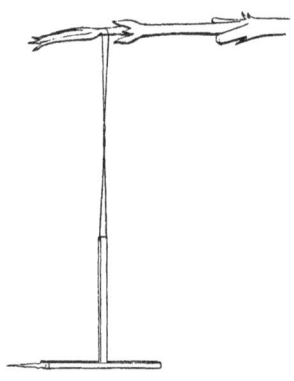

Fɪɢ. 31.   Exp. 203.

The writer (fig. 31) is made of two light pieces of wood, the horizontal piece ends in a pointed piece of platinum foil, while the vertical piece ends in a loop of cotton by which it is attached to the shoot. By twisting the loop it is easy to make the foil press lightly against the smoked paper of a revolving drum[1]; the T piece being light (about 0·2 gram) it does not interfere with the geotropic action. In a typical experiment of this sort the primary geotropic rise lasted 4 hours, the after-effect 2 hours. Then came 2 hours in which the index was stationary; finally it rose once more, beginning very slowly.

[1] We use the auxanometer drum, but a continuously rotating drum would give a better result.

SECTION B. **Curvatures due to injury, contact, etc.**

(**204**) *Decapitation of roots*[1].

Select 10 healthy germinating beans with straight roots growing vertically downwards. From 5 of them cut off 1 mm. measured from the extremity of the root-cap: the amputation must be made by a strictly transverse section, and the amputated point should include part of the growing point. Place the 10 beans horizontally in damp sawdust for 12 to 18 hours at a temperature of 15°—16° C. and compare the amount of geotropic curvature.

The result is not quite constant, it may however be safely said that decapitation prevents or greatly diminishes geotropic curvature[2].

(**205**) *Decapitation prevents the perception of the stimulus.*

Place 10 beans horizontally in damp sawdust for 1½ hours; the tips (1½ mm.) are now amputated and the roots embedded vertically in damp sawdust. After 12 hours the roots, or most of them, will be found to be curved laterally towards the side which was downwards during the period (1½ hr.) during which they were kept horizontal. This shows that amputation does not interfere with the carrying out of an induced curvature, so that the absence of geotropism in exp. 204 must be due to a disturbance of the capability of being stimulated.

---

[1] Ciesielski, *Abwärtskrümmung der Wurzel.* Inaug. Dissert. Breslau, 1871. *Power of Movement*, p. 523.

[2] A good deal of literature exists on this point and on the facts given in exp. 207. References are to be found in Frank's *Lehrbuch der Botanik*, I. p. 477.

(**205** A)   *Pfeffer's experiment.*

The final proof that the root-tip alone is sensitive to the gravitation-stimulus has been given by Pfeffer[1] and by his pupil Czapek[2]: the following instructions are taken from their publications.

As material, the authors recommend *Vicia faba* and *Lupinus*, the latter being more sensitive but also more liable to accidental curvatures.   A number of glass tubes in which the roots are to be constrained to grow into a desired form must first be prepared.   These are L-shaped, one end being closed, the other open; their total length is 3 mm. and they weigh about 30 milligrams.   Czapek says they are easily made by drawing out a thick-walled tube of soft glass in the blowpipe: the rect-angular bend is got by heating a very short region, care being taken to avoid narrowing the bore by a sudden bend.   One limb of the tube is closed in the blowpipe at 1·5 mm. from the angle, the other limb broken across at the same length, the edges being afterwards rounded in the flame.   The bore of the tube must depend on the size of the roots used, and it is important that the tubes should not fit too closely.

The caps are cemented to a sheet of cork, to which seedling beans, lupins, &c. are pinned in such a way that the tip of each root is contained in the open limb of a L tube, and reaches nearly to the bending place.   The cork plate

---

[1] British Association, August, 1894, *Annals of Botany*, VIII. Sept. 1894, p. 317.

[2] *Jahrb. f. wiss. Bot.*, XXVII. 1895, p. 243.

is now fixed to a klinostat and kept in slow rotation in damp air for 8 to 12 hours. The gravitation-stimulus being removed by the use of the klinostat, the roots grow freely into the glass tubes and are forced to assume their form: thus each root has a sharp rectangular bend at 1·5 mm. from its tip. If the experiment has been properly done the tubes should fit the roots so loosely that they can be taken off and replaced with ease: this is said to be a condition of success.

The specimens having been removed from the cork plate, they should be fixed (the tubes still attached) in various positions. If the terminal 1·5 mm. is vertical and the basal part of the root horizontal, no geotropic curve occurs, although the root grows vigorously. The part of the root within the horizontal limb of the tube is displaced by the new growth within the tube and gradually emerges from the tube. The fact that the growth of the root continues horizontally seems only explicable by the supposition that the tip of the root alone is geotropically sensitive, and since this points vertically downwards it is in a satisfied condition, without any tendency to curvature. Other specimens should be fixed with the terminal 1·5 mm. horizontal; under these conditions the vertical portion of the root outside the tube bends laterally until the tip of the root is vertical. In this experiment the cotyledons may be above, the main axis of the root pointing ver-tically downwards, or *vice versâ*, the root may be directed upwards. For the different form of the subsequent curvature in the two cases, Czapek's figs. 2 and 3 should be consulted.

(**206**) *Recovery from the effect of amputation.*

If the roots in exp. 205 (p. 173) are allowed to remain undisturbed for 3 or 4 days, the growing point is regenerated and the roots recover the power of geotropic curvature. They will grow into an S-like form, because until the growing point is regenerated they continue growing horizontally or obliquely in the direction impressed on them by the geotropic curvature induced before amputation. When the power of reacting to the gravitation-stimulus is restored they curve downwards. Fig. 32[1] shows this state of things, *L* indicates the side of the seedling which was downwards while the root was horizontally extended, *AB* shows the strong induced curvature, *BC* the second geotropic curvature occurring after regeneration of the root-cap.

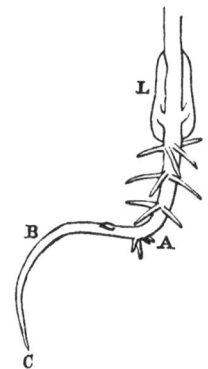

Fig. 32. Exp. 206.

(**207**) *Curvature induced by contact, injury, &c.*[2]

If the tip of a bean root is amputated by an oblique cut with a razor, so that the growing point is laterally injured, the root curves away from the injured side. The beans should be pinned to the lid of a jar and may be grown either in damp air or with the tips in water. It is important that the temperature should not exceed 16° C.

A similar curvature may be produced by contact. Minute squares (1·5 mm. × 1·5) of cardboard or of very

---

[1] *Power of Movement*, fig. 195, p. 527.
[2] *Ibid.*, Chap. III.

fine sand-paper (which adheres well) are to be fixed to the
tips of bean roots by a thin layer of strong shellac varnish
which sets hard very quickly.   It is important that the
squares of card should be attached to the slope of the root-
apex, and that they should be either on the right or left of
the root-tip : if they are on the anterior or posterior face of
the root, the resulting curvature will be in the plane of
the cotyledons and therefore liable to be interfered with
by Sachs' curvature which occurs in the same plane.   The
side on which the card must be placed will be understood
from Fig. 33, which represents bean roots with cards
attached, and showing various degrees of curvature.

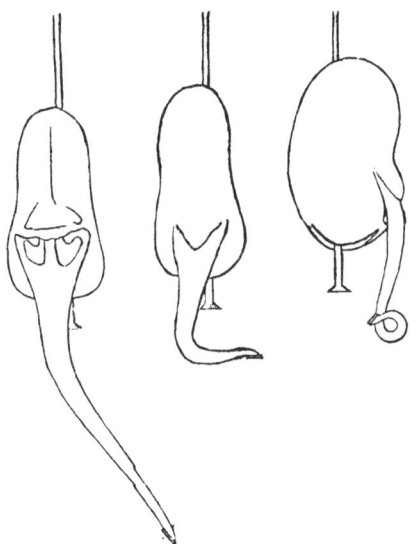

Fig. 33.   Exp. 207.
(From *The Power of Movement,* fig. 65, p. 134.)

(208) *Ciesielski's experiment*[1].

This experiment is of interest in relation to the mechanism of growth-curvatures because it shows that curvatures might conceivably arise through unequal turgescence of the two sides of a plant-member.

Take a germinating bean with rather a long root, say 40—45 mm., and let it lie on the table for a minute or two, so that it may begin to wither. Impale the seed on a pin transversely to the plane of the cotyledons, and fix the pin in a clamp so that the root is horizontal and 1—2 mm. above a surface of water. Then gently bend the root downwards till its lower surface and the water meet. In a few minutes the tip of the root will be seen to be rearing itself into the air, which is due to the increased turgescence of the lower surface of the root.

The same result may be obtained to a still more marked degree if the root is rendered flaccid by immersion in 5 % NaCl solution before being placed in contact with the surface of water.

(209) *Drooping of leaves during a frost*[2].

In connection with some of the older theories on the mechanism of growth curvature (*see* exp. 198, p. 167) it is of interest to note that flaccidity may be a cause of curvature. The leaves of the laurel, *Prunus laurocerasus*, and of the Portugal laurel, *P. lusitanica*, droop in a striking way during sharp frost. If a branch is brought into a warm

---

[1] Ciesielski's Breslau Dissertation, 1871, quoted by Sachs (*Arbeiten*, I. p. 219). See also Sachs, *loc. cit.* p. 398.

[2] Moll, *Archives Néerlandaises*, T. xv. p. 13 of separate copy.

room the leaves rapidly assume a normal position. The droop is due to a sharp curvature in the petiole: if leaves are cut and placed in lukewarm water the straightening of the petiole occurs at once and is clearly visible

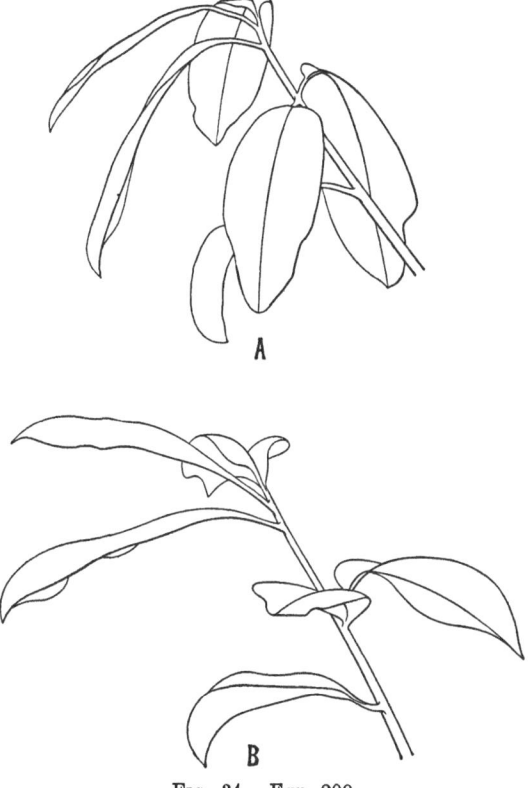

A

B

FIG. 34.   Exp. 209.

to the naked eye. Fig. 34 shows a laurel twig, A in the frozen, and B in the thawed condition. It seems

12—2

probable that these movements are due to flaccidity, because if a branch is fixed so that its axis points vertically downwards the leaves still sink during frost; in this case the sharp curve of the petiole does not take place and the position of the leaves with regard to the branch is quite different[1].

SECTION C. **Heliotropism.**

(**210**)   *Heliotropic curvature.*

Positive heliotropism may be observed with the hypocotyls of mustard (*Sinapis alba*), or with seedlings of Canary grass (*Phalaris canariensis*), which latter are extremely sensitive to small differences of illumination[2].

Sow *Phalaris* in a small pot and let the soil be level with the rim of the pot, which would otherwise shade the plants.   Place the pot on a plate of sand and cover it with an inverted cylinder of cardboard or zinc-plate, the edge of which rests in the sand, and keep it in a dark room. When the seedlings are some 10 mm. in height, remove the cylinder and let them be exposed to a small gas-flame at a distance of about 10 feet.   The light should be so faint that when the observer stands by the plants he cannot read the figures on his watch, and cannot distinguish any shadow cast by a pencil on white paper. The dark room should have walls, ceiling and floor of

---

[1] During frost the curious downward curvature of the branches of the lime (*Tilia*) should be noticed : frozen branches assume their normal form when they are brought into a warm room.   See Caspary, *Report of Internat. Hort. Exhibition*, 1866, also Geleznow, *Bull. Acad. Petersburg*, 1872.

[2] *Power of Movement*, p. 455.

a dead unvarnished black, and care should be taken that there are no polished objects which might reflect light. The room should, moreover, have double doors separated from each other by a space, so that the observer may enter the room without admitting light.

The canary grass should be left for 8 or 10 hours, when a distinct heliotropic curvature should be visible.

(211)   *After-effect.*

After-effect may be observed in the same way *mutatis mutandis* as has been described for geotropism (exp. 202).

(212)   *Light of high refrangibility most effective*[1].

To expose plants to light of different refrangibility we use a box (blackened inside) whose lateral opening can be closed by a flat bottle: the bottles may be filled with various fluids and in this way the efficacy of different parts of the spectrum may be roughly tested[2]. The bottles must fit into grooves so that no light can enter the box except through the coloured fluid.

In one box, *B*, let the bottle contain a solution of potassium-bichromate, and let the bottle in *C* contain ammoniacal copper-sulphate. In each box place a pot of *Sinapis* seedlings which have been grown in complete darkness, and whose vertical hypocotyls are about 20 mm. in length, or *Phalaris* may be used. They may be examined after 4 to 6 hours, when a striking difference should be seen between *A* and *B*.

[1] Wiesner, *Heliotropische Erscheinungen im Pflanzenreiche. Denkschr. d. k. Akad. Wien*, 1878.

[2] It is far more satisfactory, but not so easy, to use a pure spectrum.

(213) *Negative heliotropism.*

Allow mustard seed to germinate in sawdust, which should be thrown lightly into the flower-pot, not pressed down, and should not be too wet.   When the radicles are 15—20 mm. in length the seedlings are pulled out of the sawdust and placed with their roots in water.   A disc of thin cork plate is pierced with holes of 5—6 mm. in diameter and is supported by 3 bent wires in the mouth of a glass beaker about 1—2 mm. above the surface of the water.   The mustard seedlings are supported in the holes by little plugs of cotton-wool, and should be so arranged that the radicles are vertical.   The beaker is then placed near the window in a box with a lateral opening.   After a variable number of hours the roots will show a strong curvature from the light.

(214) *Struggle between the effects of light and gravitation.*

Phycomyces is strongly heliotropic as well as apogeotropic; Elfving[1] has shown that if the surface on which the spores are sown is illuminated from below by means of an oblique mirror, the hyphæ grow downwards in obedience to the stimulus of light, but in opposition to their apogeotropic tendency.   The bread on which the *Phycomyces* grows should be fixed to the inside of the cover of a stoppered jar.   A little water should be poured into the jar to keep the air damp.   The jar is then darkened, except below, by black material tied round it.   It must be suspended vertically and placed near a window, with a

[1] *Acta Societatis Sc. Fenniæ*, T. xii. 1880.

good mirror so arranged that the bread is illuminated from below. After one or two days the cover should be removed and the vertically downward growth of the mould noted.

(215)   *Transmitted stimulus*[1].

Sow *Setaria italica* in a pot in which the soil is level with the rim of the pot, and let it remain in absolute darkness for from 3 to 5 days, when the seedlings should be 12—15 mm. in height. In *Setaria* and some allied genera the seedlings are remarkable for the long hypocotyl on which the cotyledon is borne; heliotropic curvatures are produced by the bending of the hypocotyl, but the cotyledon alone is sensitive to lateral illumination, and when it is kept dark no heliotropic bend occurs, although the hypocotyl itself is lighted from one side. To darken the cotyledon small caps of tin-foil must be made in the following way. The foil is cut into squares of 8 × 8 mm. and these are rolled, each like a cigarette paper, round the base of a small pin from which the head has been cut. The pipes so made are closed by pinching them at one end with a forceps. The pinched part should be about 3 mm. in length and should be bent as well as pinched. The little hollow cylinders so constructed can be picked up with a forceps and slipped over the cotyledons of some 6—8 *Setaria* seedlings; the pot should then be exposed to lateral illumination. When time has been allowed for

[1] See *Power of Movement in Plants*, Ch. IX, for experiments on *Phalaris*, &c. The observations on *Setaria* are given by Rothert in the *Berichte d. deut. bot. Ges.*, Jahrg. X, also in his work *Ueber Heliotropismus* (Breslau), 1894.

heliotropic curvature to occur, the contrast between the capped seedlings, which remain upright, and the others, which are sharply bent towards the light, is striking.

SECTION D. **Diaheliotropism, Diageotropism, Epinasty, Nutation of Epicotyls.**

(**216**) *Diaheliotropism or transverse heliotropism* [1].

Before proceeding to experiment in this subject, it is well to study the positions naturally assumed by leaves

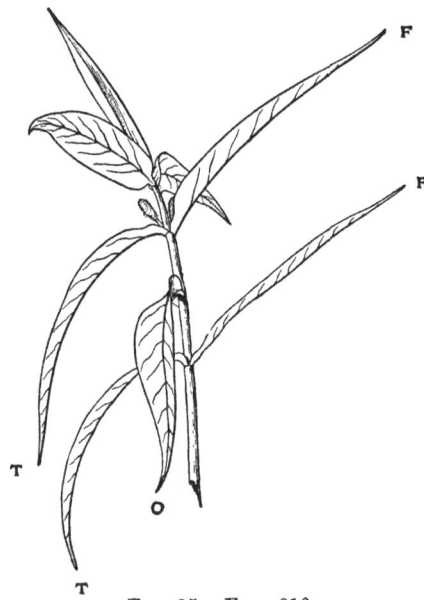

FIG. 35. Exp. 216.

when obliquely illuminated. For this purpose plants with decussate leaves are useful, e.g. *Lamium album,*

[1] Frank, *Natürliche wagerechte Richtung von Pflanzentheilen* (Leipzig), 1870.

Syringa vulgaris, and especially the shrubby *Veronicas*
such as *V. traversi* and *salicifolia.* The drawing, fig. 35,
represents the position of the leaves of the latter species,
the light being supposed to fall obliquely from the left.
It will be seen that the leaf *T* which points towards
the lighted side has a curve in its petiole which brings
the surface of the leaf at right angles to the light;
the leaf *F* is also at right angles to the light, but
it is brought into that position by the petiole being
above instead of below the horizon. The leaf *O*, which
points towards the observer[1], is twisted on its petiole and
thus reaches the position of maximum illumination by a
third movement differing from those of either *T* or *F*.
To observe the occurrence of the above movements it is
only necessary to transplant a *Lamium,* or to fix a twig of
Veronica in a bottle of water, the stem in either case
being tied to an upright stick so that no curvatures,
except those of the leaves, may occur. The angles made
by the leaves with the horizon should be noted, and
should again be measured after a few days.

(217) *The movements due to specific sensitiveness.*

It was at one time believed that the diaheliotropic
position was simply the result of a balance struck between
such opposing tendencies as apheliotropism, apogeotropism,
epinasty, &c. &c., and that diaheliotropism as a specific
form of sensitiveness was non-existent. This view has
now given way to the belief that a leaf in placing itself

[1] In the figure the leaf opposite to *O* has been removed to make the
drawing clearer.

at right angles to incident light is replying in a specific
manner to stimulation, just as a positively heliotropic
stem reacts in its specific way by placing itself parallel to
incident light.

To illustrate this fact, plants must be subjected to
one-sided illumination while kept in slow rotation on the
klinostat.

We use the instrument designed by Mr H. Darwin
and described by one of us[1] in the year 1880.

The instrument is shown in fig. 36; the plant in a

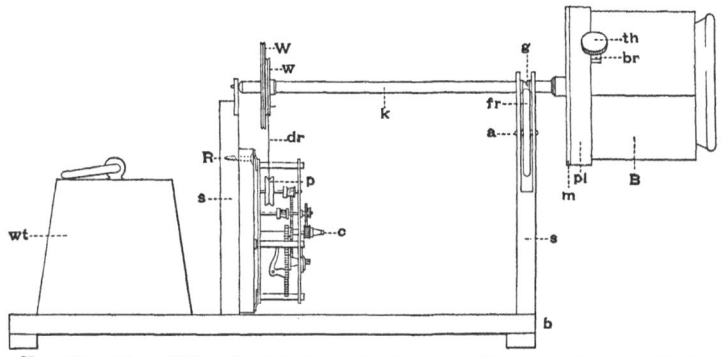

FIG. 36.   Exp. 217.   Copied from the *Linnean Society's Journal*, 1880.

flower-pot is fixed in a wooden box *B*, which again is
secured by the thumb-screw *th* to the plate *pl* : the box
being cubical can be fixed either as shown in the figure
or with the axis of the flower-pot at right angles to the
spindle (*k*) of the klinostat.   The plate *pl* is attached
to the spindle *k*, which ends in a point turning in the

---

[1] Francis Darwin, *Linnean Society's Journal*, Vol. XVIII. p. 420.   The
original klinostat is described by Sachs in his *Arbeiten*, II. p. 214.

upper end of the left-hand support *s*. The spindle is also supported at *g* on the friction wheel *fr*. The spindle (with the plant attached) is made to rotate by means of a band of silk *dr* passing round the wheel *w*, and also round a pulley on one of the axles of the American watch-action clock *c* which[1] is attached by means of the screw *R* to the support *s*. By passing the driving gear over the large pulley *W* the spindle is made to rotate once in 30 minutes. We find that one turn in 20 minutes, which is the rate given by the smaller wheel *w*, is a convenient speed.

When a plant is fixed into the box *B* it naturally happens that the centre of gravity of the plant and flower-pot does not coincide with the spindle, so that the clock has varying amounts of work to do in different parts of the rotation. The Cambridge Scientific Instrument Co.'s klinostat has an arrangement by which the position of the weight can be altered until its centre of gravity is at the centre of rotation.

The mechanism is shown in fig. 37. The end of the spindle *k* terminates in a screw *sc*, which passes through the boss *c* and the disc-shaped plate *d* to which the boss is united. As long as the end of the spindle does not press against the brass plate *n*, the disc (and spindle with it) can slide in any direction parallel to *n* in a cylindrical cavity *sp* sunk in the wood plate *pl*, of which the floor is formed by *n*, and the roof by the brass plate *m*. The latter plate is attached to the wooden plate *pl*, and is pierced by the hole *hh*. The edges of this hole limit the

---

[1] In the figure the clock has been simplified.

amount of excentric movement of the spindle. If the
screw *sc* is made to project through the disc and press
against *n*, the disc is forced against *m* and the spindle is
secured in an excentric position.

To get the weight to balance about the spindle the
following is the best plan. The screw *sc* is loosened enough
to allow the disc *d* to be moved by
the application of force, but not
enough to allow it to slip by the
weight of the plant. The plant is
allowed to take up its natural position, which will be with its heavier
side downwards: the box is then lifted
by a hand placed under it, until the
spindle no longer touches the friction
wheel, and the boss *c* (fig. 37) is struck
vertically from above with a hammer.
The spindle having been thus displaced
in the right direction, is replaced on
the friction wheel and its state of
balance is tested. The hammering must be repeated

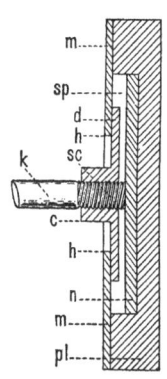

FIG. 37.   Exp. 217.
Copied from the
*Linnean Society's
Journal*, 1880.

until, when the spindle is made to rotate, it has no
marked tendency to come to rest in one position rather
than another.

The clock can be rotated in the vertical plane by
turning on the screw *R*, and thus the driving gear can be
tightened or slackened. When a new loop of silk has to be
adjusted on the wheel, or when any operation connected
with the experiment has to be performed, the clock should
always be stopped by inserting into the balance wheel one

end of a bit of lead wire, of which the other end rests on the board *b*. If this precaution is neglected the loop of silk may become entangled in the clock wheels, or the clock may be forcibly stopped by touching one of the wheels in such a way that the escapement becomes fixed; and this never happens if the balance-wheel is stopped as above described. If proper care is taken a box of earth and plant weighing together 1000 grams may easily be kept in constant rotation. If the apparatus seems at all top-heavy a heavy weight *wt*, fig. 36, may be placed on the board.

For experiments on diaheliotropism *Ranunculus ficaria* is useful, as it is obtainable early in the year and grows healthily indoors. We cultivate *Ficaria* by wrapping the roots in wet cotton-wool protected by a covering of india-rubber cloth. The plants so treated are fixed by means of large pins to a cork disc taking the place of the box *B* in fig. 36. The klinostat is placed as close as possible to the window, and it is desirable, though not necessary, to keep extraneous light away by a black curtain hung behind and at the sides of the apparatus.

The following fundamental experiments should be performed[1].

(i) When a *Ficaria* plant is dug up, its leaves freed from the resistance of the soil, curve (epinastically) strongly downwards. If such a plant is fixed with its axis horizontal and parallel to the spindle of the klinostat, and if the

---

[1] See F. Darwin, *loc. cit.*; also Vöchting, *Bot. Zeitung*, 1888; Krabbe, *Pringsheim's Jahrbücher*, Vol. xx.; Schwendener and Krabbe, *K. Preuss. Akad. Abhand.* 1892.

klinostat is placed with the spindle at right angles to the plane of the window, the leaves will be pointing almost directly away from the light. The angular divergence from the vertical of a few leaves having been noted the klinostat is set in movement. After two or three days it will be found that the leaves have curved towards the light, and that they have come to rest in such a position that the laminæ are roughly vertical, i.e. at right angles to the horizontal illumination from the window[1].

(ii) Dig up a *Ficaria* with a large ball of earth (so that the epinastic curve cannot occur), and place it in the dark, in which case the leaves will bend upwards. Leave it in the dark until the leaves are about 45° above the horizon, and fix it on the klinostat precisely as described for (i). The leaves will now be pointing more or less towards the light, and in two or three days they will have curved away from the light until their blades are approximately vertical.

(iii) In this experiment the klinostat stands as in (i) and (ii), but the plant is arranged so that its axis is at right angles to the spindle. The leaves behave in the same way as if the plant was stationary, that is to say, those pointing towards the light curve downwards, while those pointing away from the light move upwards until the laminæ of both are at right angles to incident light[2].

---

[1] Strictly speaking it is the *resultant* of the illumination which gives the effect of horizontal light. See F. Darwin, *loc. cit.* p. 427.

[2] Into the difficult question of the behaviour of the lateral leaves on the klinostat we do not propose to enter.

(**217** A)   *Exclusion of both heliotropic and geotropic curvature.*

If the klinostat is arranged so that the spindle is parallel to the plane of the window, heliotropic as well as geotropic effects are avoided. This may be simply shown by keeping a pot of seedlings rotating for a day or two, and observing that they do not curve but remain straight.

(**218**)   *Rectipetality.*

A pot of young mustard seedlings (*Sinapis*) which have been grown in the dark is placed on a klinostat placed in a darkened room. The clock is stopped and the plants allowed to remain in one position in the dark, until a measurable geotropic curvature is produced. The clock is then set in motion and after 24 to 36 hours the curvature is found to be greatly diminished or to have quite disappeared. This automatic recovery from growth-curvature has been named rectipetality by Vöchting[1].

(**219**)   *Mode of action of the klinostat.*

Two views are possible as to the mode of action of the klinostat. It might be supposed that the plant does not curve geotropically (or heliotropically as the case may be) because it never has time to perceive the stimuli. Or, it might be supposed that the stimulus is perceived but

---

[1] *Die Bewegungen der Blüthen und Früchte*, 1882.

is equally distributed. An experiment of Elfving's[1] shows
that, in some cases at least, the latter is the explanation.
If a grass-haulm which has finished growing is kept in a
vertical position, the pulvinus undergoes no change, but
growth does take place in the pulvinus of a grass-haulm
kept in slow rotation on the klinostat. This seems
to prove that just as the gravitation-stimulus acting on a
horizontal stationary pulvinus produces one-sided growth,
so an equally distributed stimulus produces a symmetrical
growth, i.e. a simple increase in length of the pulvinus.
The pulvini must be marked, and measured microscopically,
and they may then be fixed inside a large corked test-tube
and kept in rotation for 3 or 4 days. If the test-tube
contains a little water the haulms will be kept sufficiently
damp. As a control, similar haulms must be placed
vertically in a stationary test-tube. This precaution is
necessary because the pulvini may not have completed
their growth so that the control specimens will show
a certain amount of elongation.

(220)   *The peg or heel of Cucurbita.*

A similar conclusion may be drawn from the
behaviour of germinating cucurbits on the klinostat.
When a seed of *Cucurbita ovifera* (vegetable marrow) is
allowed to germinate in normal conditions the well-known
peg or heel, shown in fig. 38, *A*[2], is developed on the

---

[1] *Öfversigt af Finska Vetenskaps Societets Förhandlingar*, 1884.

[2] Fig. 36 A is from the *Power of Movement in Plants*, p. 102.

physically lower side of the radicle. But if the seeds are kept in slow rotation on the klinostat until they ger-

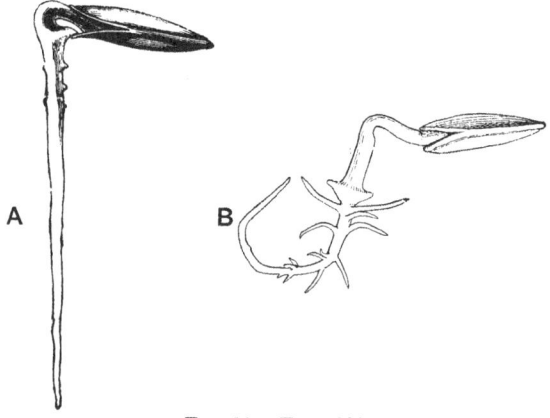

FIG. 38. Exp. 220.

minate, the peg is not developed laterally, but like a frill all round, as shown in fig. 38, *B*.

(221) *Diageotropism or transverse geotropisms*[1].

A bean, in which the secondary roots are just beginning to appear, is placed in damp sawdust in one of Sachs' trough-like glass vessels (see exp. 191). It must be placed close to the glass so that the behaviour of the secondary roots can be seen from outside. The trough is placed in the dark and the slightly oblique direction in which the secondary roots grow must be noted. If it is desired to keep a permanent record of the experiment,

---

[1] Elfving (Sachs' *Arbeiten*, II. p. 489) used the runners of *Eleocharis*, *Sparganium* and *Scirpus maritimus* for similar experiments on diageotropism.

the position of the roots should be traced on the outside of the glass with a paint-brush filled with white paint or some clearly visible bright colour, such as vermilion[1]. One end of the trough must now be raised on a wooden block, so that the edge of the trough makes an angle of about 45° or 50° with the horizon, and the curvature of the secondary roots, by which they return to their original position with regard to the horizontal, must be noted.

(**222**)   *Growth of secondary roots in light.*

If the trough is left exposed to light instead of being kept in the dark room the secondary roots grow more obliquely downwards[2]. The same thing may, according to Stahl, be observed in the runners of *Adoxa moschatellina*. This is not due to a directive influence of light, it is rather that light influences the mode of reaction to the gravitation-stimulus. A somewhat similar state of things is described in exp. 229.

(**223**)   *Diageotropic flowers.*

The horizontal position assumed by the corolla-tube of various species of *Narcissus* is due to diageotropism[3]. The movement may be easily observed in *Narcissus poeticus* by using either flowers attached to the plant or cut specimen placed in water as shown in fig. 39. Select a specimen in which the scape is vertical and the corolla-tube approximately horizontal: place it (not necessarily

---

[1] Sachs, *Collected Papers*, II. p. 885 (from the *Arbeiten*, I).

[2] Stahl, *Berichte d. deut. bot. Gesellsch.* 1884.

[3] Vöchting, *Bewegungen der Blüthen und Früchte*, 1882.

in the dark) in a position so that the corolla-tube is 45°
below the horizon. The curvature of the flower-scape will
begin to diminish and the corolla-tube will, in about 24
hours, become once more horizontal, as shown in the figure,

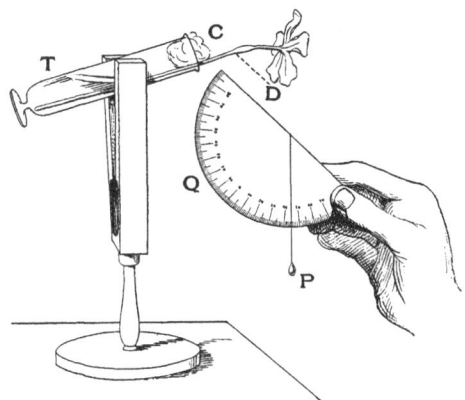

FIG. 39.   Exp. 223.

in which the dotted line D gives the position from which
the flower has risen. The angle made by the corolla with
the horizon may be roughly measured, as shown in the
figure, by means of a graduated semicircle of cardboard Q
to the centre of which a plummet P is suspended. The
zero should be in the centre of the arc and the graduations
should run from 0° to 90° in either direction.

It is of interest to note that although the curvature is
diageotropic, yet that light has a directive influence on
the flower[1]. The flower-buds of *Narcissus* are at first
directed vertically upwards, and the *direction* in which
they bend, in assuming the horizontal position, is de-

[1] Vöchting, *loc. cit.*

13—2

termined by light, although the *amount* of such movement
is regulated by the gravitation-stimulus.

(224) *Horizontal branches.*

The horizontal position assumed by some branches is
according to Frank[1] due to diageotropism. The following
observations are worth making although they leave it
undecided whether the horizontal position is due to light-
or gravitation-stimulus.

In the spring the developing buds of the hazel
(*Corylus*), hornbeam (*Carpinus*), elm (*Ulmus*), and lime
(*Tilia*) are curved so as to point downwards, and as
further development proceeds they move up into a horizontal position. Select a horizontal branch of one of the
above plants in which the terminal bud is directed
vertically downwards, and fix the branch vertically upwards so that the bud is horizontal. It will be found
that in this case the curvature of the bud remains
unchanged, so that the branch into which it develops is
at right angles to the older part of the branch with which
it is continuous.

(225) *Torsion of internodes.*

In many plants, as Frank has shown[1], the arrangement
of the leaves on the adult horizontal branches is normally
and regularly produced by torsion of the internodes. This
is the case with *Weigelia, Symphoricarpus, Philadel-
phus coronaria*, and some species of *Lonicera*. The
original decussation of the leaves is plainly seen in the

[1] Frank, *Die natürliche wagerechte Richtung*, etc., 1870.

developing buds, but when the internodal twist has taken place in a complete manner (which is not always the case) the decussation disappears, and the leaves seem to be arranged in two instead of four ranks.

(**226**) *The buds of the yew* (*Taxus*).

In the young leaf-bud of the yew it will be seen that the morphologically upper surfaces of all the leaves are directed towards the centre.   Since the shoot grows horizontally it is clear that the leaves growing on its upper surface must twist on their petioles in order that their upper surface may face upwards[1].   If a yew branch bearing quite young buds is fixed in a horizontal position with its lower surface upwards, the twisting of the leaves takes place precisely as it does in a normal branch.   Thus the shoot, although in the arrangement of its leaves it resembles a normal shoot, really developes with its morphologically upper side downwards.   In this respect it differs from the shoots of the hazel, lime, etc. in which an inverted shoot recovers its normal position by means of torsion of the internodes.   The best way of fixing the yew branch in its inverted position is not to twist it on its axis but to bend over a horizontal branch into a $\subset$ form, until its free end is inverted but horizontal; it can be fixed in this position by being tied in two places to a stick fixed into the earth.   The torsion of the leaves requires several weeks for completion.

[1] See a good figure of *Abies pectinata* in Frank's *Lehrbuch der Botanik*, 1892, I. p. 475.

(227) *Epinasty*[1].

This curvature (due to internal stimulus) may be observed in a variety of plant-members. The strong epinastic curvature of the leaves of *Ranunculus ficaria* has been made use of in exp. 217, and similar curvatures by which the leaves are pressed against the ground are to be seen in *Plantago media* and in *Pinguicula*.

(228) *Combination of epinasty and geotropism*[2].

The leaves of the dock (*Rumex*) serve for this experiment. Gather 6 or 8 leaves and with a knife free the lamina from the midrib[3], and cut off the apical third of the midribs; now fix the basal $\frac{2}{3}$ of the midribs horizontally in an embankment of wet sand, piled up in the angle of a tin box which must have a close-fitting lid. Let half the number of midribs be in the normal position, while the remainder are reversed so that the upper surface of the midrib is downwards. After 24 hours it will be found that the latter, in which apogeotropism and epinasty combine, are far more curved than the normally placed midribs.

(229) *Nutation of epicotyls.*

An interesting feature in the curvature of the epicotyls

[1] H. de Vries in Sachs' *Arbeiten*, I. p. 252; see also Vines, *Annals of Botany*, 1889.

[2] H. de Vries in Sachs' *Arbeiten*, I. p. 255.

[3] Note that the curvature of the midrib increases on being freed; this indicates a state of tension between the lamina and the midrib. See H. de Vries in Sachs' *Arbeiten*, I. p. 241.

of some of the Leguminosæ is the part played by light in the phenomenon. Wortmann[1] has shown that the epicotyls of *Phaseolus multiflorus* remain curved in the dark and straighten when exposed to light. We use the vetch (*Vicia sativa*), in which the phenomenon is particularly striking, the darkened epicotyl being often curled into a complete loop.

[1] *Bot. Zeitung*, 1882, p. 915.

CHAPTER VIII.

FURTHER EXPERIMENTS ON MOVEMENT.

SECTION A.   *Stimulus of contact, chemical agency, moisture, changes in illumination and in temperature.*

SECTION B.   *Autonomous movements: Periodicity.*

SECTION A.   **Stimulus of Contact, &c.**

(**230**)   *Tendrils: sensitive to contact.*

Among the most sensitive of tendrils are those of *Sicyos angulatus, Passiflora gracilis* and *Echinocystis lobata:* the common bryony (*Bryonia dioica*) is however more generally accessible, and being a native plant requires in ordinary summer weather no special arrangement with regard to temperature. Avoid the very young tendrils and select one with a slight hook at the end. With a pencil or rod rub the inside of the terminal part of the tendril for a minute. It will almost at once show signs of curvature, and will be strongly curved in 2 minutes,—so that for instance the terminal 15 mm. form a complete ring of 3 or 4 mm. in diameter.

(231) *Tendrils : Pfeffer's contact experiment.*

Pfeffer[1] has shown that only rough bodies which produce discontinuous pressure serve to stimulate tendrils, while perfectly smooth homogeneous bodies are not irritating.

Melt 2 sheets of Marshall's " Leaf Gelatine " in half a cupful of warm water[2] : dip into it while still hot a smooth wooden or glass rod about 3 mm. in diameter.   In this way a length of 4 or 5 cm. must be thickly coated and allowed to cool thoroughly before use.   The gelatine-coated rod, which must be kept as free from dust as possible, is now to be employed to touch the tendrils of *Bryonia* as described in exp. 230.   It will be found that the tendrils show no signs of curving ; they must now be touched with the uncoated part of the rod, to prove that the absence of movement is not due to want of sensitiveness.

On a windy day (if the experiment is made out-of-doors) it may be a little difficult to make sure that, in the first part of the experiment, the tendril does not come into contact with the uncoated part of the rod : it is for this reason that a good length of gelatine coating is recommended.   The tendril can however easily be held still by touching its convex (insensitive) side with another gelatine rod ; the slight stickiness of the gelatine fixes the tendril while the observer manipulates with the first rod.

---

[1] *Untersuchungen aus d. bot. Institut zu Tübingen,* I. 1885, p. 483.

[2] Pfeffer uses solutions containing from 5 to 14 per cent. of air-dry gelatine.

(**232**) *Mimosa : movements produced by stimulation.*

The nature of the movements of *Mimosa sensitiva* may be seen by giving the plant a shake, when the main petioles will be seen to sink, the second petioles to move together, and the leaflets to close. If the plant is healthy, and is in a moist atmosphere and at a temperature of at least 16° C., the leaves will rapidly recover their former position and may again be irritated.

An individual leaf may be made to move by gently touching the pulvinus on the under side. Note that a touch on the upper side has no such effect. The period elapsing before the leaf recovers from the effect of the touch varies with the temperature ; thus in one instance a leaf recovered in 8 min. at 23° C. ; in from 12—15 min. at 18° C.

If a lighted match is held beneath a pair of terminal leaflets they close and the passage of the stimulus may be traced by the closing of pair after pair of leaflets, until it reaches the point where the secondary petioles spring from the main leaf-stalk. The leaflets of the other secondary petioles close one after another in reverse order, i.e. beginning from the base. If the irritation is strong enough the main petiole sinks, and neighbouring leaves may also be affected.

Lastly, the whole plant may be stimulated by an irritant vapour such as that of ammonia. Place a watch-glass, containing *liquor ammoniæ fortiss.*, diluted with half its volume of water, near the plant and cover it with a bell-jar. In a few minutes movements indicating stimulation begin in the leaves nearest the watch-glass : the bell should be then removed, as the plant is easily injured by ammonia.

(**233**)   *Mimosa ; temperature.*

On the bottom of a glass cylinder place a layer of wet sawdust in which a small *Mimosa* growing in a pot may be sunk ; the cylinder is to be placed in a large inverted bell-jar filled with water of which the temperature can be varied by ice or by hot water as the case may be. The cylinder must be covered with a glass plate through a hole in which it is possible to touch the plant so as to test its sensitiveness. To get a clear result the temperature should be lowered from 20° C., at which the plant is thoroughly irritable, to 11° or 12° C., although the lower limit of irritability is about 15°. Cooling the air to this amount, by the addition of ice to the water in the bell-jar, is a tedious process, and it would probably be better to have a second bell-jar ready filled with iced water, to which the cylinder containing the plant might be transferred.

(**234**)   *Mimosa : effect of darkness.*

If *Mimosa* is kept in the dark for several days it loses its sensitiveness. The plant should be kept in a damp atmosphere in a greenhouse at a temperature of at least 16°—17° C. The best plan is to place the flower-pot in a tray of wet sawdust and to invert over it a tin cylinder, the rim of which should sink into the sawdust. We find that 4 or 5 days are needed to destroy sensitiveness. In

one of Sachs' experiments[1] a plant was placed in the dark
at 9 p.m. Sept. 24, and on Sept. 27 at 9 a.m. sensitiveness
had almost, and on 28th, quite disappeared.   It was
then exposed to light and took some days to recover.

(235)   *Mimosa: continued stimulation.*

A light stick about 25 cm. in length is transfixed by a
needle at about 6 cm. from one end; the ends of the
needle are pushed into rubber corks, which are held in a
clamp.   Since the needle cannot turn easily in the
rubber which supports its two ends, the stick will stand
out horizontally : a slight blow on the short end of the
lever makes the other end jump up, and the elasticity of
the corks brings it back to its old position ; if repeated
blows are thus applied, the long arm will make corresponding beats upwards.   These can be applied to the
leaf of the *Mimosa* by a pin fixed at the end of the
lever so as to strike the pulvinus across its longitudinal
axis.   The rhythmical blows are applied to the short end
of the lever by drops of water falling from a height of
30 cm. at the rate of 8 or 10 per minute.

With this arrangement the leaf falls at first, but
as the blows are continued the petiole rises, and after 15
minutes is insensible not only to the light blow of the
lever (otherwise it would not have risen) but also to a
more severe disturbance.

The above method is a slight modification of Pfeffer's[2].
Another plan is to tie a thread to a branch of the plant,

[1] *Physiologie* (French Trans.), p. 49.
[2] *Physiol. Untersuchungen*, 1873, p. 57.

the other end of the thread being attached to the pen-
dulum of a metronome.   In this way a rhythmical series
of shocks are applied.

(**236**)   *Oxalis acetosella.*

In the absence of *Mimosa pudica* the irritable leaves
of *Oxalis acetosella* or of some of the other trifoliate
species such as *O. stricta* or *corniculata* should be studied[1];
O. rosea is remarkably sensitive.   During the day the
three leaflets are spread out horizontally; when irritated
they sink downwards and may move through as much as
90°, though when not perfectly irritable or when not
strongly stimulated the movement is often much less.
They are not nearly so sensitive as the leaves of *Mimosa*,
and a repeated or somewhat prolonged shaking of the
flower-pot is needed to produce a good effect.   Individual
leaflets may be stimulated by rubbing the under-surface
of the pulvinus.   They recover slowly, as much as three-
quarters of an hour or an hour being sometimes necessary.
No transmission of stimulus from one leaflet to the next
has as far as we know been observed.   If leaves are placed
in alcohol in the expanded and also in the contracted
condition they retain their respective positions, according
to Pfeffer, and if median longitudinal sections are made,
the disappearance of the furrows on the upper side and
their increase on the lower surface of the pulvinus can be
seen in the contracted as compared with the expanded
leaves[2].

[1] Pfeffer's *Physiologische Untersuchungen*, 1873, p. 69.
[2] See Pfeffer, *loc. cit.* p. 70 : see also his figures 5 and 6.

(237)   *Oxalis : Brücke's experiment*[1].

Oxalis acetosella provides convenient material for the repetition of Brücke's classical experiment, by which it may be shown that the rigidity of the pulvinus diminishes after stimulation.   Pfeffer's method[2] is to fix the stalk of the leaf in a bottle of water by means of a bored cork, which is cut into a cone at its upper end so that the cork may come close up to the pulvinus and yet allow the leaflets room to fall.   We prefer to cement the petiole to a vertical pin fixed into an ordinary flat-topped cork.   A needle-point to act as an index is fixed at the heart-shaped extremity of the leaflet so as to form a continuation of the midrib.   According to our observations the needle should be fairly heavy or the excursions of the leaflet are not large enough.   A C-shaped piece is cut out of card, which is made to serve as a graduated arc, and it is fixed to the bottle so that the position of the leaflet can be recorded by means of the graduations.   The first reading must be taken when the leaf is in its normal position, the bottle being held so that the leaflet is horizontal.   The bottle is now turned over so that the leaflet is still horizontal but upside down, and a second reading is taken.   If the pulvinus were absolutely rigid the first and second readings would coincide ; as it is they differ by 5°—15°.   The leaf being replaced in the first position, the pulvinus must be irritated by rubbing it on the lower surface, or by a blow or two on the leaflet.

---

[1] Müller's *Archiv für Anatomie und Physiologie*, 1848, p. 434.
[2] *loc. cit.* p. 74.

After one or two minutes have elapsed to give time for
the leaflet to sink, the above process of reading—reversing—and reading again must be gone through, when the
angular movement will be considerably increased, and
may be even double what it was before stimulation. The
same specimen should be used to measure the rigidity of
the leaflet in the nocturnal position. It will be found
that the excursion is very much smaller at night,—something like $\frac{1}{4}$ of what it is in the day.

(**238**)   *Drosera: stimulated by meat.*

The readiest way to observe the general character of
the movement of the tentacles is to place a very minute
fragment of raw meat on the gland of one of the outer
ones; this usually causes strong inflection in 7 or 8
minutes, by which the meat is after a time carried to the
centre of the leaf. The gland should be carefully watched
under a lens in order that the time may be noted which
elapses between stimulation and the beginning of the
movement. An active leaf ought to show distinct movement in half-a-minute. In all experiments on *Drosera*
leaves should be selected which have good drops of
secretion on the glands, and which show a healthy red
colour. Leaves which are too old are of a *dark* red and
should be avoided.

(**239**)   *Drosera: irritated by contact with inorganic matter.*

The tentacles may also be excited by other substances,
e.g. by fragments of pounded glass placed on the glands.
It is well to use coloured glass, so that it may be possible
to see easily whether the fragments actually touch the

glands or only rest in the secretion.   In the latter case
no movement occurs.

The glands are not sensitive to a single touch : the
impact must be rapidly made, but may be hard enough to
bend the tentacle.   Nor has the blow caused by falling
drops of water any stimulating effect.

But water containing even the finest solid particles
produces movement.   This may be shown by placing half-
a-dozen leaves in distilled water rendered milky by a
little precipitated chalk, an equal number of leaves being
placed in pure distilled water[1] for comparison.   In ten
minutes the leaves in the emulsion should be well in-
flected.

In Darwin's *Insectivorous Plants* a number of experiments are given which show that excessively light bodies
resting on the glands cause inflection.   Pfeffer has however proved[2], when sufficient care is taken to prevent the
vibration of the room reaching the leaf, that a smooth
particle of glass which is comparatively heavy may rest
on the gland without producing movement of the
tentacle.

(240)   *Drosera : inflection caused by dilute solutions*[3].

Very dilute solutions of ammonium phosphate cause
inflection.   Prepare with good distilled water a solution
of ammonium phosphate in the proportion of 1 to 50,000
(2 centigrams to a liter).   Fill 10 watch-glasses with the

---

[1] Carefully distilled water should be used, and even this sometimes
causes after a time a certain amount of inflection.

[2] *Untersuchungen aus d. bot. Institut zu Tübingen*, I. p. 513.

[3] *Insectivorous Plants*, p. 153.

solution and 10 others with the distilled water used in making the solution. In each glass place a healthy leaf, and examine them after an hour, when the phosphate leaves should show much inflection, whereas only a few isolated tentacles in the control leaves ought to have moved. A longer period than one hour may be required in some cases.

(**241**) *Drosera: asymmetrical inflection.*

If a particle of raw meat, or preferably a minute fragment of calcium phosphate, is placed in the middle of a leaf the exterior tentacles all bend towards the centre. If however the object is excentrically placed, e.g. halfway between the centre and circumference of the disc of the leaf, the tentacles no longer bend symmetrically towards the centre, but are plainly directed towards the phosphate[1].

(**242**) *Berberis (Mahonia) aquifolium: irritable stamens*[2].

The flowers do not easily lose their irritability; cut twigs in water may be used, or even isolated flowers, if only moderate care is used to prevent them withering. Some of the larger flowered species are more convenient to work with than those of *B. aquifolium*.

In the condition of repose the anthers lie in the bifid hoods of the petals : when the filaments are irritated they curve inwards, bringing the anthers close up to the stigma. To localise the irritable part, the anther should

---

[1] See *Insectivorous Plants*, Fig. 10, p. 244.

[2] Heckel, *Comptes rendus*, 1874, Vol. 78.

first be gently stroked on both faces with a dissecting needle or mounted bristle. No movement occurs, but the filament springs in at once when touched on its inner face just below the anther. The outer surface of the filament is not sensitive except at its extreme base. To make sure of this the petals should be dissected off, an operation which requires a little care, and the specimen so prepared should be placed on wet filter-paper under a watch-glass for 10 minutes to recover irritability. Under a simple lens it is now easy to touch the filament in any part.

The filaments apparently begin to recover from the effect of a touch at once, at any rate a considerable amount of return towards the resting position is visible in 1 or 2 minutes, and in 5 or 10 minutes recovery is complete.

The stamens may be irritated separately, no transmission from one to the next takes place. By applying the tetanising current they may be made to close simultaneously: we simply wrap one wire round a small twig of *Berberis* and touch the stigmas with the other wire. Or a single flower may be cut off and held by one of the wires stuck into the pistil, the other wire being applied to the base of the flower outside.

(**243**) *Berberis: effect of chloroform.*

Gather a flower carefully with a pair of forceps, test its irritability by touching a single filament and place it floating in a watch-glass of water. Add a couple more flowers similarly treated and place the watch-glass under a bell (1000 c.c. capacity) with another watch-glass containing 4 or 5 drops of chloroform. The flowers can withstand 10 minutes of this atmosphere without suffering and will be found quite insensible to touches.

After half-an-hour's exposure to fresh air (and possibly in a shorter time) they are found to be once more irritable.

It is interesting to repeat the experiment with flowers in which the filaments have been irritated just before they are put under the bell-jar. It will be seen that the stamens recover the normal position—in spite of the anæsthetic[1].

(244) *Stigma of Mimulus cardinalis.*

The stigma has the form of a pair of divergent lamellæ which, when irritated, rapidly shut together so that one stigmatic surface meets and presses against the other. The inner surface is the sensitive part, a touch on the outside of the lamellæ produces no effect. Oliver[2] has shown that if one lamella is prevented from moving, a touch on it still provokes movement in the other lamella. In our experiments we fixed one lobe by cementing it to the corolla, using mastic dissolved in ether for the purpose. When the cement is dry it is well to push a strip of wood between the style and the corolla: by a wedge of this

---

[1] The same thing is said to occur in the case of *Mimosa* : see Pfeffer, *Physiologische Untersuchungen*, 1873, p. 64.

[2] *Berichte d. deutschen botan. Gesellsch.* v. 1887, p. 167. Transmission of stimulus has only been observed in *Martynia lutea, M. proboscidea* and *Mimulus cardinalis*. There is said to be no transmission in *Mimulus luteus.*

sort the fixed lamella is forced to be more or less horizontal and is perfectly free from contact, even at the base, with the opposite lobe.

(245) *Stamens of Centaurea cyanus.*

We use the garden form of this species for demonstrating the fact that the stamens are irritable[1].

Select a floret which has expanded but has not extruded its style, place it on a piece of wet filter-paper and under a simple lens split the floret at the swollen part of the corolla-tube, which can then be opened wide enough to give a view of the filaments. Cover the preparation with an inverted watch-glass and leave it for 15 minutes,—to recover from the effects of the operation. If the filaments are now gently touched, a writhing contraction is plainly seen. The anther tube may in this way be made to swing over first to one side, then to another, as the filaments are made to contract on the corresponding sides.

To see the extrusion of pollen another flower must be used : the best plan is to remove the pollen at the free end of the anther-tube with a camel-hair brush. If this is done with a rapid touch, the filaments are irritated and a ribbon of pollen emerges immediately after the blow.

(246) *Phycomyces: curvature towards iron.*

Elfving has shown that the sporangiferous hyphæ of *Phycomyces nitens* curve towards iron. It is only necessary

---

[1] Pfeffer employed *C. jacea* and *Cynara scolymus*. See his *Physiologische Untersuchungen*, 1873, p. 80.

to plant an iron rod in the centre of a *Phycomyces* culture
(from which light is carefully excluded), and leave it for
12 or 18 hours, when the hyphæ are seen bending from all
sides towards the rod. The cause of this remarkable
phenomenon is still obscure[1].

(**247**) *Hydrotropism.*

The curvature of roots towards a moist surface can be
demonstrated by the well-known method of Sachs[2]. A
sieve is constructed by fastening netting to a bottomless
box or stretching and tying it over the mouth of a short,
wide[3] cylinder open at both ends and made of galvanised
iron or tin-plate. A thin layer of moist sawdust or finely
divided cocoa-fibre is spread on the netting, seeds are
placed on the layer and covered with 2 inches of the
same material. The sieve is now hung up so that the
bottom makes an angle of about 50° with the horizon.
As the roots emerge they leave the vertical and grow
along the moist surface of the sieve. We find that
cereals such as rye or barley answer well. The only
difficulty is to provide a suitable atmosphere in which to
hang up the sieve. If the air is too dry the roots wither
before they have time to bend; if too damp the surface of
the sieve does not supply a sufficiently strong contrast to
the surrounding air. A greenhouse atmosphere answers
fairly well, or, as Sachs recommends, a large cupboard or

---

[1] See however Elfving's interesting paper in *Nature*, March 15, 1894,
where references to his original paper and to Errera's work on the subject
are given.

[2] Sachs' *Arbeiten*, I. p. 212, Fig. 3.

[3] 5 cm. deep, 20 cm. in diameter.

dark room of which the floor is occasionally watered. According to the same authority the air should be in such a condition that the difference between the wet- and dry-bulb thermometers is $1\cdot5°$—$2\cdot0°$ R. ($2°$—$2\cdot5°$ C.). We find it a good plan occasionally to squirt with water the lower surface of the sieve.

(**248**) *Movement of chloroplasts.*

We find that the leaves of *Oxalis acetosella*[1] give good results. Ten or twelve leaves are taken from a plant, and after the stalks have been cut short off beneath the pulvini they are placed floating in water. Half of the number in one dish are exposed to bright sunshine, the rest remain in dull diffused light. After two hours they may be examined by preparing surface sections of the spongy parenchyma. In the sunned leaves the chloroplasts are in the "profile position," that is, they lie against the side walls of the stellate parenchyma cells, and may even be crowded into the corners. In the shaded leaves they are spread out and dotted over the surfaces which are parallel to the plane of the leaf. The leaves may be preserved in alcohol for future examination.

(**249**) *Chemotaxis: antherozoids.*

The following instructions are taken from Pfeffer's paper in his *Untersuchungen*[2]. The prothalli which yield the antherozoids for Pfeffer's experiments were chiefly

[1] This plant is recommended by Stahl, *Botanische Zeitung*, 1880. Stahl's figure of *Oxalis* is copied in Frank's *Lehrbuch*, p. 289.

[2] *Untersuchungen aus dem botanischen Institut zu Tübingen*, I. 1881—1885, p. 363.

small ones of *Blechnum fraxineum* and *Adiantum cuneatum*[1].
They were grown on lumps of peat in the shade of other
plants and were used when only about a millimeter in
length, and had numerous antheridia but few or no
archegonia. They should be kept only moderately damp,
as this seems to favour the yield of antherozoids. The
prothalli having been washed for a moment are placed 3
or 4 together under a small cover-glass supported on
strips of paper, and are washed by repeatedly drawing a
current of rain water through the preparation. Distilled
water, being injurious to the antherozoids, must not be
used for the washing, the object of which is to remove any
malic acid which may be set free by the rupture or injury
of the tissues of the prothallus. The reasons for
preferring a small cover-glass are that the antherozoids
are thus confined to a smaller space, and that the water is
better oxygenated than when a large glass is used.
Capillary tubes of 0·1 to 0·14 mm. internal diameter and
from 7 to 12 mm. in length are closed at one end by
melting the glass, and are filled with the malic acid
solution with the help of an air-pump. The solution may
be either the free acid or a salt, for instance, sodium
malate ; the solutions should be made with rain water and
contain about 0·05 per cent.[2] A capillary tube is pushed
under the cover-glass, when the antherozoids in the
neighbourhood of the opening are at once attracted to it.
Pfeffer has seen 60 antherozoids enter a tube of malic

---

[1] The young prothalli of *Ceratopteris*, grown from spores sown on
bricks, give a good supply of antherozoids.

[2] The strength may vary from 0·01 to 0·5 per cent.

acid within half-a-minute from the beginning of the experiment.

(249 A) *Chemotaxis: Bacteria[1].*

Allow a boiled pea to decay in about 200 cc. of water for two or three days: draw off some of the fluid from just below the surface, transfer a drop to a slide and place on it a small cover-glass raised on two strips of paper. Capillary tubes, like those used in exp. 249, are made by drawing out coarse tubes in the blowpipe flame and again drawing out the tubes so made over a small flame. Lengths (10 mm.) of the fine capillary tubes should be sealed, each at one end, and filled with 2 p.c. KNO_3 under the air-pump. They may be removed from the beaker of KNO_3 with a camel-hair paint-brush, and should be examined under the microscope to make sure that they are full. Having been washed with a drop of distilled water they should be pushed under the cover-glass. In about 10 minutes the open ends ought to be crowded with Bacteria.

(249 B) *Chemotaxis: pollen tubes[2].*

This may be easily demonstrated with the wild hyacinth (*Scilla nutans*). A thin jelly is made by adding 3 p.c. of good gelatine (for instance the " bacteriological " gelatine sold by Baird and Tatlock) to water and warming over a water-bath. A large drop is placed on a slide and

[1] Pfeffer, *Untersuchungen aus dem bot. Institut zu Tübingen*, II. 1888, p. 582.

[2] Molisch, *Sitzb. d. k. Akad. d. Wiss. in Wien*, II. 1893; and Miyoshi, *Flora*, 1894, p. 76.

into this is stirred a quantity of pollen. Then the stigma and one or two ovules are inserted at various places in the jelly while it is still fluid. Care must be taken to avoid air-bubbles. The slide is then placed in a saturated atmosphere, preferably in darkness. After a few hours, in warm weather, the pollen tubes will be seen under the microscope directed towards the stigma, the cut-end of the style and the ovules.

Plantago media is also a good plant for the experiment, and various species of *Reseda* work well if 1 p.c. of cane-sugar be added to the gelatine.

The gelatine soon becomes mouldy and fresh material must be made up, or else the old must be kept sterile by repeatedly heating to 80° C. in a water-bath.

The most perfect demonstration of chemotaxis may be made by mounting *Scilla* pollen with a stigma in a drop of 10 p.c. cane-sugar on a slide, and arranging a damp chamber round it with blotting-paper and a cover-slip. The slide is arranged on the microscope and must not be shaken or moved. After a few hours the cover-slip is taken off and the slide examined. The pollen of *Narcissus Tazetta* may be used in the same way in 7 p.c. sugar.

(249 c) *Chemotaxis : pollen tubes.*

A preparation of pollen in jelly is made as described in exp. 249 B, except that the ovules and stigmas are omitted and that a cover-glass is placed on the drop. The pollen tubes will be observed to grow away from the edge of the cover-slip towards the centre of the drop, i.e. from places rich in oxygen to places poor in oxygen. Cane-sugar

solution may also be used, of various strengths for
different plants[1], e.g. *Fritillaria imperialis*, 15 p.c.;
Narcissus Tazetta, 7 p.c.; *Vincetoxicum officinale*, 15 p.c.
It should be noted[2] that all pollen tubes do not exhibit
this phenomenon.

(250)   *Opening and closing of the tulip: temperature.*

Many flowers open with a rise of temperature and
close with a fall; the best adapted for experiment are the
crocus and tulip[3]. Both of these are sensitive to slight
changes of temperature, and both are valuable because
they can be made by appropriate treatment to open and
shut at any time of the day. The crocus is the more sensitive of the two, but the tulip answers extremely well, and
the following instructions apply to this genus.

It is convenient to begin the experiment on a cool,
cloudy morning, when the tulips are naturally closed.
Cut a flower and fix it vertically in a cork fitted into a
bottle of water. To one of the outer perianth segments
and to the opposite inner segment fix filaments of glass
drawn out to very fine capillary tubes. They are best
cemented with shellac varnish to the groove or line running down the centre of the outer surface of the segment.
The filaments, each of which projects 3 cm. beyond the
flower, serve as indices for noting the movements of the
segments. The simplest plan is to fix a millimeter scale
horizontally so that the distance between the points of
the indices can be read off. The tulip should be prepared

---

[1] See Molisch, *loc. cit.*, where a list of suitable solutions is given.

[2] See Molisch, *loc. cit.*

[3] Pfeffer, *Physiologische Untersuchungen*, 1873, p. 181.

in a room free from sunshine, and where the temperature is not above 15° C.,—a temperature of 11° or 12° better still.

The flower having been left to itself for 15 minutes is placed in a temperature of about 20° C. In 5 or 10 minutes a clear increase in the reading on the scale shows that the flower is opening.

It may now be replaced in a temperature of 10°—12° C. Notice that the flower continues to open for some time and then begins to close. The same phenomenon *mutatis mutandis* is to be seen on changing a low into a high temperature. It is easy to make a tulip open, close, and open again within one hour.

(**251**)   *Tulip: sensitiveness to small change of temperature.*

Pfeffer[1] has seen a crocus flower open slightly in 15 minutes during which the temperature rose by less than 1° C. The change of temperature was produced by opening the door between a cold and a warm room. For class-work it is perhaps best to try rather larger changes of temperature. A tulip, fitted with two indices as described above, shows distinct opening in half-an-hour when moved from a temperature of 13·5 C. to a temperature of 15·5°, closing slightly again on being replaced in a temperature of 13° C.

(**252**)   *Crocus: mechanism of the movement.*

The following instructions are based entirely on Pfeffer's[2] account of the experiments, in which he showed

---

[1] *Physiologische Untersuchungen*, 1873, p. 183.

[2] Ibid., p. 167.

that when a crocus or tulip opens it does so because of
the accelerated growth on the inner faces of the segments,
and *vice versâ* when it closes. A series of 4 or 5 minute
dots (about 1·5 mm. apart) are made with black spirit-
varnish on both surfaces along the region of curvature,
which in the crocus is the lower $\frac{1}{4}$ or $\frac{1}{5}$ of the perianth-
segment. The distance between the marks must be
measured with great care by means of an eye-piece
micrometer: the amounts of growth observed do not
exceed 3 p.c., and it is therefore necessary to use a
magnifying power of something like × 80, and a micrometer with which the distance between the marks on the
flower is about 200 divisions of the micrometer. To get
accurate measurements it is necessary to sketch each of
the varnish marks, noting on the drawing a corner or
projecting point from which the reading is taken. The
readings are easily taken on the outside of the perianth
segment, and by removing the opposite segments the
inner marks can also be observed. The readings are
assumed to have been taken on a closed flower, which is
then placed in a room warmer than the first by 6°—7° C.
and after $\frac{1}{4}$ hr., during which the flower opens, the readings
are again taken. On the inner side the marked region
will have increased by 2·5 p.c., on the outer side an
increase of say 0·2 p.c. will be noted. If the readings are
taken first in the open flower and then in closed condition, precisely the reverse is noted, namely, that the
inner side increases only a little, while the outer side
grows about ten times as much.

(253) *Light and darkness: Bellis.*

Among the flowers which close in darkness and open when illuminated the daisy (*Bellis perennis*) is the most universally accessible[1].

A daisy should be cut and fixed vertically in a bottle of water, when the position of the ligulate florets must be noted; this may either be done by taking the angle which the flower-head fills when looked at in profile, or by measuring the horizontal distance between the tips of two opposite florets. In one of our experiments a daisy was darkened at 2 p.m., and the angle showed a diminution of 30° by 3·15.

Further experiments on the daisy are given in the next section.

(254) *Light and darkness: Trifolium.*

Sleeping plants can be made to assume the nocturnal position by darkening them in the daytime. This can be shown in any of the common species of clover, such as *T. repens*. The simplest plan is to cover a plant (growing in the open air) with an inverted vessel made of opaque material, scattering dry powdered soil round the outside of the rim so as to make sure that light is excluded. After one or at most two hours the plants may be examined, when the leaflets will be found in the nyctitropic position shown in fig. 40, where the left-hand leaf is awake, the one on the right asleep: the lateral leaflets are face to face and the terminal leaflet folded on

[1] Pfeffer, *Physiologische Untersuchungen*, p. 198.

to their edges.   The experiment may also be made with
a sod of clover dug up and kept wet in a basin or even
with cut leaves in a bottle of water.

FIG. 40.   Exp. 254.
From *The Power of Movement in Plants.*

By covering up one plant with a hollow bell-jar
containing potassium bichromate, and another with a bell
containing ammoniacal copper sulphate, it may be shown
that the orange light acts like darkness, while the blue
acts like daylight.

Finally, a few plants should be kept dark for 5 or 6
days to observe the fact that the leaflets ultimately
assume a position resembling the day-position, except that
the leaflets droop somewhat.

(**255**)   *Nyctitropic movements.*

To get a general idea of the varied character of
nyctitropic movements it is best to compare the diurnal
and nocturnal positions in a selection of plants.

Trifolium has already been described ; as a contrast it
is well to examine the nocturnal positions of a trifoliate
Oxalis, such as *O. acetosella* (in which the leaflets point
nearly vertically downwards at night), and of *Marsilea
quadrifoliata* (in which the four leaflets rise and arrange
themselves in a vertical packet).   The nyctitropism of

Melilotus, with its curious right and left-handedness[1], of
Cassia, in which the leaves sink and twist, and of *Desmodium gyrans*, in which the vertical droop of the larger
leaflets is particularly striking, should also be studied.
Movements not produced by means of a pulvinus, but by
the growth of the leaf-stalk should be examined; for
instance the nocturnal rise of the young leaves of *Nicotiana glauca* or of the cotyledons of the cabbage and
radish (*Brassica oleracea* and *Raphanus sativus*).

In all these cases note that the nocturnal is more
nearly vertical than the diurnal position, and that when
there is close contact between neighbouring leaflets it is
generally the upper surface of the leaf that is protected.

(**256**) *Nyctitropic movements : Mimosa.*

In order to study the sleep movements of leaves more

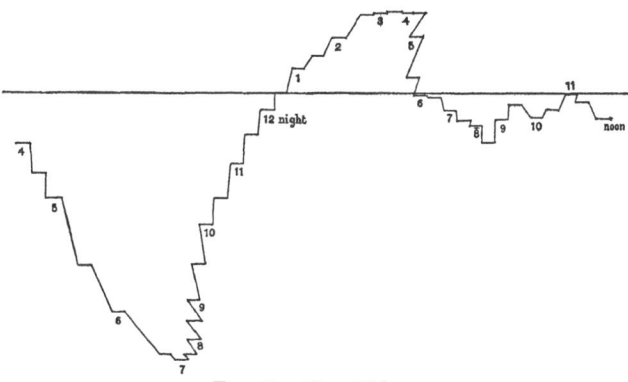

Fig. 41.   Exp. 256.

closely we employ a self-recording method. Fig. 41 is

---

[1] *Power of Movement in Plants*, p. 346, fig. 140.

a copy of a tracing[1] made by the main petiole of *Mimosa pudica* from 4 p.m. Aug. 16, until noon of the following day. The tracing was made by means of the hanging writer described in experiment 203, which recorded the position of the petiole at intervals of half-an-hour on the revolving drum used for auxanometer experiments. The tracing only records changes in the vertical position of the free end of the petiole, and does not give the angle which the petiole makes with the horizon, but if a few readings of the angle are taken, the rest can be calculated from the known length of the petiole and writer. It will suffice for our present purpose to know that at 7 p.m. the petiole was roughly 15° below, and at 4 a.m. 60° above the horizon.

The tracing shows that the leaf sank with increasing and then decreasing rapidity from 4 p.m. to 7 p.m., when it rose (at first slowly) until 3 a.m. It then remained stationary until 4·30, when a fall again occurred, followed by irregular movements continuing to noon.

(**257**)   *Paraheliotropism : Averrhoa bilimbi.*

The leaves of many plants assume in bright sunshine a more or less vertical position, which has been sometimes called " diurnal sleep" but is now known as paraheliotropism. *Oxalis acetosella,* in which the leaves in bright sun assume the same vertically dependent position that they take at night, is a familiar example. *Averrhoa bilimbi*

---

[1] To diminish the horizontal extension of the diagram, the horizontal lines drawn by the writing index are reduced. The engraving is moreover reduced by ⅓ from the drawing so prepared.

(one of the Oxalidæ) also drops its leaflets in sunshine as it does at night. The leaves of *Averrhoa*, as described in exp. 261, exhibit remarkable autonomous movements[1], in which the leaflets drop rapidly through 15°—20°, then rise slowly to their original position, repeating the movement once in 15 minutes or so. When sunshine strikes the plant the leaflets fall until they make an angle of 70° or 80° below the horizon, but this is not effected in a single drop, but by series of rapid rises and falls as

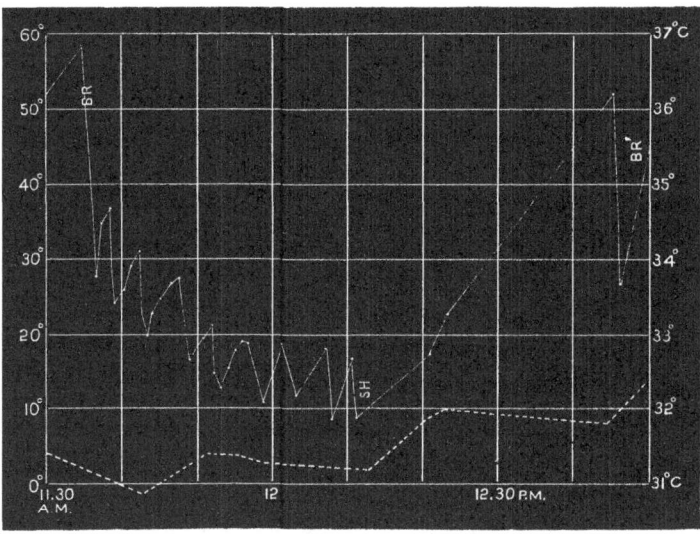

FIG. 42. Exp. 257. From *Power of Movement*.

represented in fig. 42. In this diagram the numbers 0° to 60° on the left represent the angular divergence in

[1] Lynch in *Linnean Soc. Journal*, XVI. p. 231 ; *Power of Movement in Plants*, p. 330.

degrees of the leaflet from the vertical, those on the right
represent temperature ($C°$). Thus at 11·30 the leaflet
made an angle of 52° with the vertical, i.e. an angle of 38°
below the horizon while the temperature (the dotted line) was
31·4° C. The leaflet had been slowly rising for 25 minutes,
and at BR a blind was pulled up so that the plant was
brightly illuminated, when the leaf descended in 5 steps
to the paraheliotropic position, where it executed three
rapid movements (at about 80° below the horizon), until
at SH, when the blind was pulled down, it rose for 35
minutes, to be again disturbed by sunshine at BR'. In
performing this experiment the windows must be opened
when the blinds are pulled up, to equalise the temperature
as much as possible.

The observations here recorded were made as follows.
The main petiole of the leaf observed pointed straight at
the observer, being separated from him by a vertical pane
of glass. The petiole was fixed so that the pulvinus of
one of the lateral leaflets was at the centre of a graduated
arc placed close behind the leaflet. A fine glass filament,
attached to the leaflet and projecting like a continuation
of the midrib, served as an index. As the leaflet rose and
fell its angular movement was recorded, by reading at
short intervals of time the position of the index on the
arc. To avoid errors of parallax the readings were taken
by looking through a small ring painted on the vertical
glass in a line with the pulvinus and the centre of the
arc.

SECTION B. **Autonomous Movements : Periodicity.**

(258) *Circumnutation.*

A really good method of observing circumnutation (where the movement is small) has yet to be devised. One of the methods described in Darwin's *Power of Movement in Plants* (p. 7) is, in spite of certain faults, perhaps the best for our present purpose.

Any rapidly growing plant will serve for observation, for instance, a seedling cabbage or sunflower. The most essential precaution is that the plant shall not be subjected to lateral illumination, to insure which the experiment ought to be conducted in a room lighted from above. If this is not possible the plant must be in a cylinder blackened inside, and should be illuminated by an oblique mirror hung above the mouth of the cylinder so as to throw the light vertically downwards. The plant should if possible rest on a steady stone floor.

To the upper end of the hypocotyl a delicate glass filament, 20 mm. in length, is fixed vertically by shellac varnish, which should be thick enough to dry rapidly. Before it is fixed the following preparations are necessary. A minute equilateral triangle of paper (2 or 3 mm. to the side) is pierced in the centre and is slipped over the glass filament, pushed down to the base and there fixed with shellac, so that it is at right angles to the filament. At the other end of the filament a minute bead of black sealing-wax is fixed. A sheet of glass (2 ft. × 2 ft.) fixed horizontally about 2 ft. above the apex of the plant serves as a medium on which to record the

15—5

movement.　The head of the observer is moved until the
globule of sealing-wax is exactly in the centre of the
paper triangle, and a dot is made on the glass in line

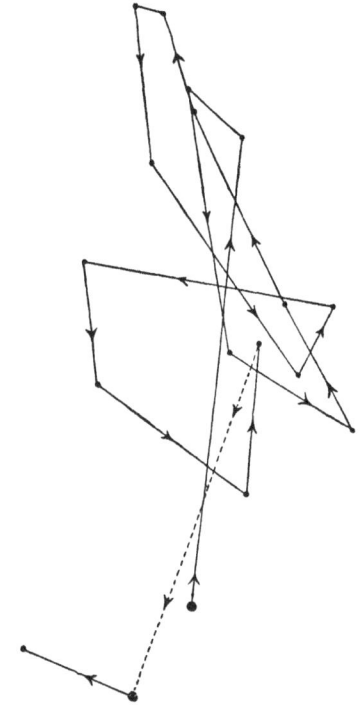

Fig. 43.　Exp. 258.　From *Power of Movement.*

with these two points.　The dot should be made with a
piece of hard wood cut to a sharp point like a pencil, and
dipped in Indian ink or moistened with water-colour.　It
is best to hold the pointer as close as possible to the
glass until the observer has made up his mind where

the dot is to be made, and then to bring the pointer sharply down on the glass. A little practice is needed to get results with small amounts of movement. The observations should at first be made at intervals of about 10 minutes, so that the observer may get an idea of the rapidity with which the movement under observation is proceeding. He may thus be able to regulate the intervals between his subsequent observations so as not to spend unnecessary time, and yet not to fail in getting a fair idea of the movement.

Fig. 43[1] represents the circumnutation of a cabbage seedling, from 9.15 a.m. to 8.30 a.m. on the following day, the dots represent the actual marks made on the glass with the sharpened wood, the lines and arrows being added to show the course of the movement: the woodcut is reduced to half the size of the original. No observations were made at night: the broken line represents the change of position which took place between the first evening and the following morning. The tracing therefore (if the broken line be neglected) practically represents the circumnutation during a day of about 12 hours.

(**259**) *Circumnutation : twining plants*[2].

The observations should be made either in a greenhouse or indoors, on *Humulus lupulus* (the hop) and *Phaseolus multiflorus*. The basal part of the plant should be tied to a stick stuck in the soil of the pot, and 6 inches or a foot

---

[1] The tracing was made by a slightly different method to that here described.

[2] C. Darwin, *Climbing Plants*, Chapter I.

of the stem should project beyond the upper end of the stick and hang over so as to be more or less horizontal. If the flower-pot stands on a sheet of paper it is easy to note, by means of a line drawn radially from the pot as a centre, the direction in which the nutating shoot points at any moment; and thus the rate at which it swings round can be recorded. One revolution in 2 hours is what may be expected in a vigorous plant. Note that the hop travels with the hands of a watch, *Phaseolus* against it : also, that if they are allowed to climb up sticks they do so by an apparent continuation of their revolving nutation. Thus a hop which has wound spirally round a support makes a left-handed screw, while *Phaseolus* is right-handed.

(**260**)   *Autonomous movements : Trifolium.*

The spontaneous *variation-movements* of leaves may easily be studied in the genus *Trifolium*[1]. They may be observed by a modification of the method described in exp. 258, or by Pfeffer's method, which is simpler and on the whole gives more reliable results. Transplant a lump of turf, containing a plant of clover, to a flower-pot : fix a vertical stick (about as thick as a pencil) firmly into the soil and attach the petiole of a leaf to it by two bands of gummed paper so that the top of the petiole is level with the top of the stick. The terminal leaflet is free to move, and its movements are observed by fixing with shellac-varnish a very fine glass filament along the midrib on the upper surface so as to project 2 or 3 mm. Cut out

[1] See Sachs' *Physiologie* (French Trans.), p. 515.

a C-shaped piece of cardboard and graduate the inner edge into 5°, i.e. make 36 graduations to the semicircle. Now attach the card to a second stick fixed in the soil so that the pulvinus is in the centre of the arc, and so that the glass index can travel over the graduations. The experiment is preferably conducted with illumination from above, but even with a side light the movements are clearly seen.

The following table gives the readings in an experiment of this sort:

Time a.m.	above horizon	Time p.m.	above horizon
Apr. 17, 8.57	27°	Apr. 17, 12.44	9°
9.43	14	1.30	14
10.18	9	3.30	19
11.45	18	4.40	15
		5.20	23

The leaf made 3 complete oscillations in 8 hrs. 23 m.

(261) *Autonomous movements: Averrhoa.*

If *Averrhoa bilimbi* is kept at a sufficiently high temperature (e.g. 27° C.), the movements of the leaflets are easily seen. Each leaflet moves independently of the rest, it falls suddenly from, e.g. 20° below the horizon to 40° below, and slowly rises in about 20 minutes to its former position. If the temperature is increased to 31°—32° C. the oscillations become more rapid and smaller in amplitude, at the same time the mean position of the leaflet falls to something like 50° below the horizon instead of 30° as at first. These changes are graphically represented in the *Power of Movement in Plants*, p. 334, fig. 135. The ordinates represent the angle made with the vertical by the leaflet under observation; a fall in the curve thus represents a drop of the leaflet, 0° representing a vertically dependent position. The dotted line represents temperature, and is to be read in connection with the numbers 76° F....90° F. on the right hand of the diagram.

(262) *Autonomous movements: Desmodium gyrans.*

The leaf of *Desmodium gyrans* represented in fig. 44 consists of one large and a pair of very minute leaflets. It is these which execute the movements for which the plant is celebrated. The point of each leaflet describes a rough sort of circle or ellipse, but the movement being exceedingly jerky and irregular the circular course is not obvious. The movement does not occur unless the temperature[1] is about 22° C., and may be remarkably stimulated by a higher temperature, thus at 40° C. the circles are made at the rate of about two in three minutes. The experiment may be performed by immersing a leaf in cool water which is gradually heated.

FIG. 44. Exp. 262.
From *the Power of Movement.*

[1] This is only true of full-grown plants. Seedlings move at lower temperature : see *Power of Movement in Plants*, p. 362.

(**263**)   *Periodicity : Bellis (light and darkness).*

Three or four among a patch of daisies on a lawn are to be darkened by covering them with an inverted flower-pot : the hole in the pot must be plugged and a layer of earth placed over the plug to make sure that no light enters. For the same reason a ring of earth should be placed round the junction of the rim of the pot with the ground. If the daisy is kept darkened for two days it will cease or almost cease to open or shut, the flower-heads taking on a permanently half-shut condition. This shows that the alternation of light and darkness is necessary for continuance of " sleep " movement.

If, however, the plants are covered in the evening the flowers will open the following morning, showing that a certain inherent periodicity exists.

If the plants are covered in the morning the periodic movement will probably begin to be irregular by the next day. For instance the waking movement will only occur late and the closure at night will also be irregular.

(**264**)   *Periodicity : Bellis (temperature).*

The flower-heads of the daisy are to some extent influenced by changes of temperature, but they behave very differently from the crocus or tulip. When they are naturally closed in the evening a rise of 15° C. in tempera-ture does not open them ; nor does a corresponding fall close them in the morning. But, according to Pfeffer, if they are warmed in the morning or cooled in the evening by about 15° an opening or closing (as the case may be) is produced[1].

The experiment in which the temperature is raised is the only one which we have confirmed. It is simply necessary to gather 3 or 4 closed daisies in the morning, place their stalks in water and put the bottle near the fire in a warm room : similar flowers being placed for comparison in a cool place out of doors.

(265) *Contrast: Bellis.*

If the flower-heads of the daisy or dandelion are kept shut throughout the day by exposing them to a temperature of 2° to 3° C. they may, according to Pfeffer, be made to open in the evening by bringing them into a temperature of 17°—20° C.[2]

A similar experiment may be made in another way. Daisies kept in the dark for two days are brought into a warm and lighted room in the evening together with control specimens growing under natural conditions. Both sets may be gathered and placed with the stalks in water. In our experiments the temperature out-of-doors was 11° C., in-doors 21° C., rising to 24° C. The daisies from the dark opened wide in 20 minutes. The control daisies showed no opening in 20 minutes, and had hardly opened after 2 hours. The degree of opening was however simply noted by means of rough sketches.

[1] *Physiologische Untersuchungen*, 1873, p. 195.
[2] Ibid. p. 197.

PART II.

CHEMISTRY OF METABOLISM.

CHAPTER IX.

Introduction.

The practical study of the transformations which
plastic substances undergo in metabolism is an application
of organic chemistry : the immediate problem is generally
to determine whether certain substances are present or
absent, and, if present, in what amounts, in particular
tissues.

For the qualitative testing of insoluble substances, such
as proteids, starch, etc. and of soluble substances which
give well-marked colour reactions, such as phloroglucin,
inulin, etc. microchemical methods are invaluable ; but for
quantitative work, and often for the satisfactory identification of certain compounds, it is necessary to make extracts
with appropriate solvents and to submit these extracts to
a systematic examination.

It is the object of the physiologist to employ the
simplest methods which will give accurate results as
regards the compounds to which attention is being given,

but it is not often that extracts can be prepared which will contain only those compounds to which the attention is directed, consideration may therefore necessarily be given to substances occurring in the extracts, although in themselves comparatively unimportant.

The necessity for quantitative results in experiments on metabolism is obvious, and even for qualitative work it is sometimes necessary to employ rather complicated chemical methods.

The arrangement followed in these sections is based on practical convenience, those substances which are commonly extracted together being placed in the same section. Each chapter contains a short general explanation of the methods to be used, followed by instructions for performing the qualitative and quantitative experiments selected.

Attention is chiefly directed to substances which are either themselves 'plastic' or are believed to have important significance in metabolic processes—the study of other compounds is only introduced in so far as these are liable to interfere with the examination of the above.

It is seldom necessary to attempt a complete analysis of all the constituents of a tissue, but it is essential to have due regard to those which may interfere with the recognition or estimation of a particular compound (e.g. tannins in the detection and estimation of sugars).

Before beginning the chemical examination of a vegetable tissue, it is of the greatest assistance to consider carefully what substances are likely to be present: a knowledge of the general distribution of the commoner plastic substances should suggest the method of procedure, but knowledge of this kind must not be relied on to the extent of substituting it for qualitative examination.

For descriptions of the practical details of many of the quantitative estimations, reference to standard works on chemistry has been freely used. A full description is given here only in cases where processes are required which are not in general use, or where there is much difference of opinion as to the best method of operating.

The principles on which the methods of estimation are based are generally explained in each case; where references only are given, care should be taken to understand exactly the reasons for all the steps described in text-books; otherwise those who have not received an adequate chemical training are liable to use instructions of this kind in a merely mechanical way, and thus to lose all the value of the work as an introduction to the practical study of this branch of physiology,—where the success of an operator largely depends on his ability to modify methods to meet particular cases.

Frequent references will be found in the text to the following works.

SUTTON. *Volumetric Analysis.* 5th edit. (Churchill, 1886.)

FRESENIUS. *Qualitative Analysis.* 10th English edition. (Churchill, 1887.)

FRESENIUS. *Quantitative Analysis.* 7th (or 6th) English edition. (Churchill, 1876, 7th.)

BEILSTEIN. *Handbuch der organischen Chemie.* 2nd or 3rd edition.

Among general works which may be consulted for descriptions of the apparatus and manipulation used in the experiments are:

SACHSSE. *Die Chemie und Physiologie der Farbstoffe, Kohlenhydrate und Proteinsubstanzen.* (Leipzig, 1877.)

FRANKLAND. *Agricultural Chemical Analysis.* (Macmillan and Co., 1889.)

DRAGENDORFF. *Plant Analysis.* (Trans. by GREENISH.) (Baillière, Tindall and Co., 1884.)

In references to original papers the ordinary abbreviations are used.

Preparation of the material to be examined.

For extraction with benzene, ether, etc., the material must be dried as completely as possible, and as the residue will generally be treated with boiling alcohol before extraction with cold water, the original substance may be dried at 100° C. in the steam-oven, till it ceases to lose weight. Fixed oils and fats, glucosides, tannins, carbohydrates including starch, will not be seriously altered by drying at 100°.

On the other hand proteids and ferments would be completely changed by drying at 100°, and in these cases either the undried material is used or material which has been dried at a temperature not exceeding 30°—this is commonly spoken of as air-dried material.

Where comparative experiments only are being made it is not a matter of much consequence whether the percentages are calculated for material which is fresh, or has been air-dried, or dried at 100°, but if determinations of several constituents are made with differently treated portions of the original substance, it is better to calculate all results in p.c. of material dried at 100°, i.e. in p.c. of dry weight.

The moisture (i.e. loss on drying at 100°) having been determined, the calculation from one state to the other involves very little trouble.

Before treatment with any solvent the substance should always be as completely disintegrated as possible —this is a point which requires very careful attention and too much importance cannot be given to it.

Ordinary dry tissues, such as leaves, portions of herbaceous stems, etc. can generally be reduced to a fine powder without much trouble, but hard tissues are often difficult to extract satisfactorily. Unless very tough— which will seldom be the case—grinding in a small mill, such as is used for grinding coffee berries, will generally bring the substance into a suitable condition.

Fresh tissues are more troublesome to work with : they can be treated in a large mortar with a small quantity of the solvent, and thus rubbed up into a perfectly homogeneous paste. The addition of some sand, or very finely powdered glass, facilitates the process and is seldom objectionable : indeed its use, in preventing the solid material 'caking' during extraction, often more than counterbalances the inconvenient increase in bulk which it causes. If it is desirable to weigh the residue at the end of the successive extractions an exactly weighed quantity of sand or glass can be used and its weight deducted from that of the total residue.

Preparation of extracts.

Non-nitrogenous plastic substances.

For non-nitrogenous plastic substances a given portion of the original material should be treated with successive solvents in the following order: (1) ether or benzene, (2) alcohol of ·85 specific gravity, (3) cold water, (4) dilute acid.

In each case the extraction must be continued till a fresh portion of the solvent fails to extract anything more.

10 grs. of material and about 250 c.c. of solvent will generally be found convenient quantities.

I. Extract the dry substance with boiling ether, benzene, or petroleum ether (with boiling point not above 75°C).

Extract (No. I.) contains oils and fats, ethereal salts of organic acids (esters), resins, terpenes, chlorophyll and colouring matters, etc.

This extraction is best performed with Soxhlet's apparatus for fat-extraction.

The material in the inner tube is weighed before and after the experiment, dried at the same temperature in each case.

When the extraction is complete, the inner tube may be placed in a steam-oven to dry the residue before weighing.

II. Extract the dry residue from I. with boiling alcohol ·85 specific gravity (about 55 p.c.).

Extract (No. II.) contains tannins, glucosides, and part of the sugars, etc.

This extraction may also be performed with Soxhlet's apparatus so that after the final weighing in I. it is only necessary to place the inner tube in another apparatus. When nothing more can be extracted by the solvent, the inner tube is again dried in the steam oven and weighed.

III. Extract the dry residue from II. with cold water.

Extract (No. III.) contains dextrins and soluble carbo-hydrates not extracted by ·85 alcohol.

The residue from II. is put into a stoppered bottle with a portion of solvent and shaken for some time in a continuous-agitation machine. The liquid is completely decanted off and a fresh portion of solvent having been added the process is repeated.

The residue is finally filtered off and thoroughly washed, the washings being added to the extract. The residue need not be dried but is ready at once for No. IV.

[A bottle fixed on to a vertical wheel which is rotated by a band from a laboratory turbine answers very well for this purpose. A rather slow rotation of about 12—15 revolutions per minute is the most effective. An appa-ratus for this purpose made to work with Rabe's turbine is sold by Gallenkamp and Co., London. A simple machine for continuous agitation is also made by the Scientific Instrument Company, Cambridge.]

IV. Extract the residue from III. with 1 p.c. sulphuric acid at 100°.

Extract (No. IV.) contains the products of the action of dilute acid on starch (dextrins and reducing sugars).

16—2

The residue from III. is placed in a flask fitted with a reflux condenser, acid added, and heated on a water-bath for two hours, or until a drop of the solution ceases to give any iodine reaction.

Instead of using dilute acid it is often better to use solution of diastase (active malt-extract) to extract the starch from the residue No. III. (see Chap. XIV.).

The temperature for extracting with diastase must not exceed 60° C. and the extraction takes considerably longer—it is best to allow the action to continue (with addition of more malt-extract from time to time if necessary) till the solution ceases to give any well-marked iodine reaction.

It will generally be found sufficient to use 100 c.c. of an extract, made by completely exhausting 300 grs. of malt with water and making up the product to 500 c.c. for each 10 grs. of substance originally taken.

It is not generally required to examine further the residue from IV., but if it is desired to estimate the cellulose in it this may be treated in the manner described on p. 293.

The original drying at 100° C. and treatment with boiling ether or benzene and boiling ·85 alcohol tend to alter the proteids and render them insoluble in cold water and dilute acids, so that it frequently happens that the extracts III. and IV. are almost completely free from proteids.

Preparation of extracts.
Nitrogenous plastic substances.

For nitrogenous plastic substances (proteids, amides,

etc.) another portion of the original material which has been dried at a temperature not above 30° C. is extracted (1) with cold water, (2) the residue from (1) with dilute alkali, 1—2 p.c. soda (NaOH).

If much oil or fatty matter is present this should be first removed by agitation with benzene in the cold; the residue washed with ether and dried below 30° can then be treated with (1) water, (2) dilute alkali.

Extract 1. (Cold water) contains soluble proteids, peptones and albumoses, amides, nitrates and nitrites, ammonium compounds, etc. etc.

Extract 2. (Dilute alkali) contains proteids insoluble in water but soluble in dilute alkali.

These extractions (and the agitation with cold benzene) may be made with the apparatus used for the cold water extracts of non-nitrogenous substances (see p. 243).

Filtration.

It is often very difficult to filter clear the extracts of vegetable tissues; the use of some of the special kinds of paper made by Schleicher and Schüll will frequently give a clear filtrate where ordinary filter-paper fails, but filtering through asbestos is very effective in troublesome cases.

For filtering with asbestos it is necessary to use a filter-pump, and the most convenient method of working is to use a perforated porcelain filter plate with a hardened paper (such as no. 575 in Schleicher and Schüll's catalogue), on to which the asbestos is poured whilst suction is applied below, the asbestos having previously been stirred into a cream with warm water; after the water has drained away the felt-like layer of asbestos can be pressed with a pestle to any degree of tightness required. The asbestos and paper can be washed and used again.

Evaporation of solutions.

Much time is always necessary for this tedious operation, but work can generally be so arranged that it can go on whilst other experiments are in progress.

Evaporations should always be conducted on the water-bath, but even so a considerable amount of 'charring' usually occurs, and this discoloration of the solutions, not being easily removable, renders the subsequent examination much more difficult.

By concentrating solutions under reduced pressure the difficulty is largely avoided. By the use of a good water filter-pump aqueous solutions can be concentrated fairly rapidly at 50°—60° C. and alcohol can be distilled off below 50° C.

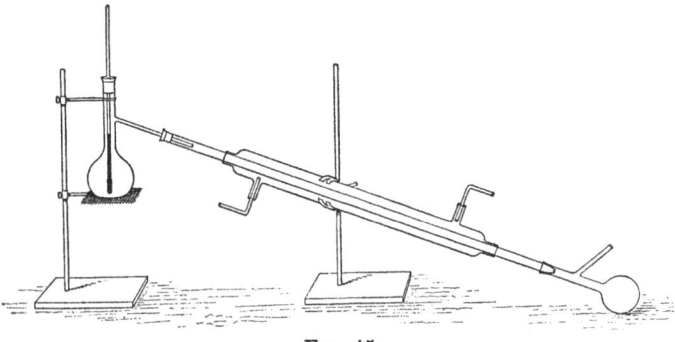

Fig. 45.

A convenient form of apparatus for distilling off liquids

under reduced pressure can be readily constructed with distilling flasks and a Liebig's condenser. The diagram Fig. 45 shows such an apparatus with the parts connected ready for distilling. Care must be taken that all the connections are air-tight.

Changes occurring in solutions on keeping.

It is necessary that the examination of the solutions obtained by dissolving in water the residue from the alcohol extract (II.), by extracting with cold water (III.), and dilute acid (IV.), etc., should be examined as soon as they are prepared.

Such solutions undergo change very rapidly from the action of micro-organisms, and in a few days will often be full of fungoid growths. If the solutions have to be put aside before they can be examined, some strongly antiseptic substance must be added to prevent such growth. Chloroform is very convenient, as it can be readily expelled by warming before the examination is made. The addition of 1 c.c. of chloroform per liter of extract will generally be found sufficient.

Small quantities of thymol may also be used for this purpose—the small amount of thymol necessary will not interfere with the subsequent examination, but its use is not applicable where any part of the solution is to be used for fermentation (see sugars).

If the alcoholic or cold-water extract is found to be acid in reaction when first prepared it should at once be exactly neutralised, because heating with acids may cause changes in some of the constituents (e.g. cane-sugar may be inverted by citric acid). Dilute soda is most convenient for this purpose.

When lead acetate, normal or basic, mercuric chloride, etc., are used in excess for precipitating, and the metal is removed from the filtrate by H_2S, free acid (acetic, hydrochloric, etc.) is produced in the solution. This should be neutralised as soon as the H_2S has been expelled by warming [acetic and hydrochloric acids are not removed by heating dilute solutions].

CHAPTER X.

PROTEIDS, AMIDES, AMMONIA, NITRATES, ETC.

FULL particulars of the chemistry of vegetable proteids, amides, etc., may be obtained from the following works.

RITTHAUSEN. *Die Eiweisskörper der Getreidearten.* Bonn, 1872.

SCHWARZ. *Die morphologische und chemische Zusammensetzung des Protoplasmas.* Breslau, 1887. (Cohn's *Beiträge,* Vol. v.)

Read papers by

PALLADIN. *Ber. d. d. bot. Ges.* Vols. VI. and VII. (1888—1889.)

E. SCHULZE. *Landw. Vers.-Stat.* Vol. XXXVI. (1889) and *Landw. Jahrb.* XXI. (1892.)

VINES. *Journal of Physiology.* Vol. III. (1881.) *Proc. Roy. Soc.* No. 191. (1878.)

GREEN. *Phil. Trans.* Vol. 178. (1887.)

SERNO. (Distribution of Nitrates.) *Landw. Jahrb.* XVIII. (1889.)

OSBORNE. *Amer. Chem. J.* XIV. (1892.)

For convenience of practical study we may consider nitrogenous plastic substances as divisible into

(1) *Proteids insoluble in water but soluble in dilute 1—2 p.c. alkali.* [e.g. the gluten of cereals.]

(2) *Proteids soluble in water.*

Substances which give all the characteristic proteid reactions and are precipitated by potassium ferrocyanide and acetic acid, trichloracetic acid, cupric acetate, and normal lead acetate, etc., etc.

They may or may not be precipitated by boiling after addition of acetic acid, or by adding excess of 90 p. c. alcohol. [e.g. the 'soluble proteids' of cereals and leguminous seeds.]

(3) *Peptones and albumoses.*

Soluble in water but not precipitated by any of above reagents : give a characteristic reddish tint with the biuret reaction : are completely precipitated from neutral solutions by alcoholic mercuric chloride, and from slightly acid solutions by sodium phosphotungstate; are slightly diffusible through membranes. [e.g. vegetable peptones and albumoses of leguminous seeds.]

(4) *Amides.*

Amido-derivatives of organic acids. [e.g. asparagin, glutamin, betain, etc.]

(5) *Ammonia, nitrates and nitrites.*

The examination of these constituents is made on special extracts of original material obtained as described on p. 244 (which are referred to below as the alkali solution and the water solution) and not on the extracts used for non-nitrogenous substances.

Proteids insoluble in water, soluble in dilute alkali.

A portion of the alkali solution is carefully neutralized with dilute acid, a precipitate (soluble in excess of acid) is formed if proteids are present.   This precipitate should be filtered, washed, and portions of it tested by the ordinary reactions for proteids (xantho-proteic, Millon's, biuret, etc.)

Proteids soluble in water.

Portions of the solution should be tested by boiling after addition of a drop of acetic acid and by the addition of 90 p.c. alcohol—if precipitates are produced they should be filtered off, washed and tested for proteid reactions.

Whether precipitates are caused or not by the above add to fresh portions of the solution :—

(1)   Potassium ferrocyanide and a drop of acetic acid,

(2)   Aqueous solution of trichloracetic acid.

Both these reagents give precipitates with proteids and they will frequently cause precipitates when the solution does not change on boiling or on addition of alcohol.

If proteids are present add cupric acetate (aqueous solution) as long as it causes a precipitate, and filter.

Peptones and Albumoses.

Remove excess of copper from the filtrate by H_2S—warm to expel excess of H_2S and concentrate the liquid if necessary.

Test portions of this solution for peptones and albumoses by:—

(1) Biuret test [an equal volume of strong soda (NaOH) solution and adding one or two drops only of dilute copper sulphate solution].

Peptones and albumoses give a characteristic colour of a redder tint than biuret.

(2) Sodium phosphotungstate and dilute sulphuric acid.

Peptones and albumoses give a white precipitate.

(3) Saturated alcoholic solution of mercuric chloride.

Peptones and albumoses give a white precipitate insoluble in water when once thrown down.

Amides.

If peptones and albumoses are present, remove by adding alcoholic mercuric chloride as long as it causes a precipitate—filter—evaporate off alcohol from filtrate and remove excess of mercury from solution by H_2S. After warming to remove H_2S, exactly neutralize solution with dilute soda and test portions for amides by:—

(1) Addition of freshly precipitated and well washed cupric hydroxide.

Amides form a deep blue liquid with solution of the hydroxide—if this liquid is carefully evaporated and allowed to stand (preferably in vacuo over sulphuric acid), characteristic crystals of copper oxide compound of amide may be obtained.

(2) A well-cooled mixture of potassium nitrite and dilute sulphuric acid.

Amides evolve nitrogen.

(3) Boiling for some time with dilute acid.

Amides give ammonia in solution which can be tested for in the usual ways, best by heating with excess of magnesia and testing for evolution of the gas.

Ammonia, Nitrates, Nitrites, may be tested for by ordinary methods in separate portions of the aqueous solution.

Ammonia, see Dragendorff, § 97, p. 81.

Nitrates⎫
Nitrites⎭, see Fresenius, *Qualitative Analysis*, 10th ed. pp. 228, 230.

The most useful reagents for these substances are metaphenylene-diamine and diphenylamine. Brucine in strong sulphuric acid, and starch solution with zinc iodide and acetic acid are also useful.

If it is required to examine further the nature of the proteids etc., this may be done by decomposing the precipitates caused by addition of metallic salts. Precipitates obtained from addition of copper, lead, and mercury salts may be decomposed by H_2S, those from sodium phosphotungstate by excess of baryta $(Ba(OH)_2)$ and subsequent removal of excess of $Ba(OH)_2$ in filtrate by current of CO_2.

Amides can be recrystallised from dilute (50 p. c.) alcohol.

Estimation of Proteids.

The methods based on weighing the precipitated proteids are seldom satisfactory, as the precipitates are generally impure. The quantity of proteid in the precipitate can most conveniently be estimated by determining the nitrogen present and multiplying by the appropriate factor for the method employed.

The estimation of nitrogen may be made by any of the methods used in organic analysis, but Kjeldahl's method and Wanklyn's albuminoid ammonia process possess the great advantage that the precipitate or residue does not require to be powdered or intimately mixed with a solid, a process always difficult in such cases and frequently almost impossible.

The soda-lime method[1] of Will and Varentrap much used at one time for the determination of nitrogen in organic residues, has been largely replaced by Kjeldahl's process.

For the estimation of peptones and albumoses, the nitrogen in the dried sodium phosphotungstate precipitate may be determined by Kjeldahl's process; or the precipitate may be decomposed by alkali and the peptones etc. in solution estimated by Wanklyn's method; or by comparison of the colour obtained in biuret reaction with the colour given by standard peptone solutions under

[1] A full account of the soda-lime process is given in Fresenius, *Quantitative Analysis*, 7th ed. vol. II. pt. I.

similar conditions. Since it is not common to find any considerable quantities of these substances in vegetable tissues under normal circumstances the colorimetric method will generally suffice.

Estimation of Proteids in alkaline solution.

[As the other nitrogenous substances have been extracted by the previous treatment with water, all the nitrogen in this solution may be taken to be in the form of proteids.]

Evaporate to dryness a portion of the alkaline solution, weigh the residue and determine the nitrogen, in the whole or a weighed portion of it, by Kjeldahl's method.

For a full account of all necessary details of manipulation and precautions desirable see Sutton, *Volumetric Analysis*, 5th ed. pp. 68—70.

Nitrogen found × 6·3 = Proteids.
  or Ammonia × 5·2 = Proteids.

Or

Add so much of the solution as will contain not more than 5—10 milligrams of proteids to 500 c.c. of pure distilled water, free from ammonia, in a large retort, and distil after addition of 50 c.c. of alkaline permanganate solution.

Determine the ammonia in the distillate by Nesslerising.

For a full account of the details of the process see Wanklyn, *Water-analysis* (Trübner and Co.), or Sutton, *Volumetric Analysis*, pp. 389 and 397.

[The second 50 c.c. of distillate should be Nesslerised first and if it is found to contain much ammonia, the first

50 c.c. of distillate should be diluted to 500 c.c. before
being Nesslerised. The first 50 c.c., if Nesslerised without dilution, will often yield a precipitate or colour too
deep to be accurately compared with the standard.]

It is absolutely necessary that all the apparatus and
water employed should be quite free from ammonia.
These operations are best performed in a room kept for
the purpose where the apparatus and pure distilled water
can be stored.

If 600 c.c. of water nearly free from ammonia are
placed in the retort before beginning the determinations
and 100 c.c. are distilled off and thrown away, the 500 c.c.
of water will be quite free from ammonia and the whole
apparatus perfectly clean.

The rapidity with which a number of estimations
can be made by this process renders it very suitable
for comparative experiments involving determination of
proteids.

$$\text{Ammonia} \times 10 = \text{Proteids.}$$

Estimation of soluble Proteids.

The nitrogen in the precipitate caused by copper
acetate may be determined by using Kjeldahl's method on
the dry precipitate, or, the precipitate may be suspended
in water, decomposed by H_2S and the proteids estimated
in solution (after expelling H_2S) by the albuminoid
ammonia process.

Estimation of Peptones and Albumoses is seldom
required, but methods by which it can be made are mentioned above, and the details can easily be worked out

from a knowledge of the processes described for proteids. For the colorimetric method it is necessary to make standard solutions of pure peptone: any good commercial peptone may be used for this purpose, but it should be carefully dried before weighing out the quantities needed for the standard solutions.

Estimation of Amides.

Amides are estimated in a solution from which proteids and peptones and albumoses have been removed by the processes described for qualitative testing. The separation of amides, if several are present, is a matter of much difficulty; it is generally sufficient to estimate the total 'amides' together and calculate the result as asparagin; results obtained in this way should be stated as 'amides calculated as asparagin.'

Sachsse's method.

This method (decomposing amides with potassium nitrite and sulphuric acid, measuring the volume of nitrogen evolved) gives very good results but is rather difficult to manipulate and, unless great care is taken at all stages of the operation, gives results seriously too high.

For the details of Sachsse's method see Dragendorff, § 241, p. 245.

With asparagin the reaction is

$$C_4H_8N_2O_3 + 2HNO_2 = C_4H_6O_5 + 4N + 2H_2O.$$

Therefore in this process 56 grs. nitrogen = 132 grs. anhydrous asparagin.

Method based on the action of dilute acids.

The solution is boiled for about one hour with 5 p.c. sulphuric acid in a flask with reflux condenser, and the ammonia produced is estimated either gasometrically by sodium hypobromite in a nitrometer, or by distillation with magnesia into a measured volume of standard acid.

With asparagin the reaction is

$$C_4H_8N_2O_3 + H_2O = C_4H_7NO_4 + NH_3.$$

Therefore in this process 14 grs. nitrogen = 132 grs. anhydrous asparagin (or 17 grs. ammonia = 132 grs. asparagin).

For details of the estimation of 'combined ammonia,' see Sutton, *Volumetric Analysis*, pp. 59 and 482.

[The gasometric process in the nitrometer is carried out as in the estimation of urea by sodium hypobromite.]

Estimation of Nitrates and Nitrites.

(α) Evaporate 100 c.c. to 5 c.c. (about) on the water-bath.

Decompose in the nitrometer with strong sulphuric acid over mercury and measure the volume of nitric oxide. This gives the NO from nitrites and nitrates [reduce volume to 0° and 760 mm.].

[See Sutton, pp. 226 and 362.]

(β) In another portion of original solution (diluted with pure distilled water if much nitrite is present), estimate the nitrite by Griess' colorimetric method.

Calculate the vol. of NO from nitrite (at 0° and 760 mm.) in 100 c.c. of original, and subtract this from the total NO observed in (a): the difference is NO from Nitrate.

[See Sutton, p. 366.]

(1 c.c. of NO (at 0° and 760 mm.)

$$= \cdot00170 \text{ grs. } N_2O_3 \text{ or } \cdot00242 \text{ grs. } N_2O_5).$$

Ammonia and other nitrogenous compounds will not seriously interfere with these reactions.

For estimation of ammonia present in aqueous solution see Dragendorff, § 97, p. 81.

Experiments on nitrogenous Metabolism.

Qualitative.

Make an aqueous extract of young plants of *Onobrychis sativa*[1] which have been grown in the dark and extract the residue (after exhaustion with water) with 2 p.c. NaOH (see p. 245).

Examine the NaOH extract for proteids insoluble in water.

Examine the aqueous extract for soluble proteids, peptones and albumoses, amides.

Examine another portion of the aqueous extract for ammonia, nitrates, and nitrites.

Quantitative.

Compare the amounts of insoluble proteids, soluble proteids, peptones and albumoses (if qualitative examina-

---

[1] The seeds should be put to germinate 8—10 weeks before the material for these experiments is required.

tion has shewn these to be present) and amides in extracts (made as for qualitative testing) of

(1)   seeds of *Onobrychis sativa*,

(2)   young shoots of *Onobrychis sativa* grown under ordinary conditions,

(3)   young shoots of *Onobrychis sativa* which have been grown in the dark or kept in the dark for several days before extracting.

Compare the amounts of ammonia, nitrates and nitrites in

(1)   normal young shoots of *Onobrychis sativa* which have grown in sand,

(2)   in normal young shoots of *Onobrychis sativa* which have been kept for several days soon after germination in the dark in absence of free oxygen,

(3)   in normal young shoots of *Onobrychis sativa* which have been grown in sand watered with a solution containing ammonium nitrate (·5 p.c.).

CHAPTER XI.

PARTICULARS of the chemistry of vegetable oils and fats may be obtained from

Spon's *Encyclopaedia*, 'Oils and Fatty Substances.'

BEILSTEIN. *Handbuch der organischen Chemie*, 3rd ed. vol. I.; Pflanzenfette.

KÖNIG. *Chemie der menschl. Nahrungs- und Genussmittel*, Berlin, 1889.

Read :—

GREEN. *Proc. Roy. Soc.* XLVIII. (1890).

SIGMUND. *Sitzb. k. Akad. Wien* (1890).

MÜLLER. *Ber. d. d. bot. Ges.* VIII. (1890).

SUROZ. (Abst.) *Bot. Cent.*, Beiheft I. (1891).

Oils and Fats.

These substances will all be in the benzene or petroleum ether extract—the extract may also contain some volatile oils, terpenes, resins, etc., but except in the case of barks the amount of these is generally inconsiderable. The oils and fats will always be mixtures of glycerides of fatty acids or of free fatty acids or of both.

It is not necessary to attempt the separation of the acids, which is a difficult and complicated operation; it will suffice to determine the amount of substances which can be saponified by potash, and the glycerin produced in saponification.

Oils and Fats.

Qualitative examination (of benzene extract).

Distil off the greater portion of the solvent, transfer the residue to a dish and completely evaporate the remainder of the solvent on a water bath or in a steam oven.

Note the character of the residue, whether liquid, solid, or semi-solid, etc.

Warm the residue on a water-bath with strong potash solution for about one hour, dilute with water and filter if necessary (the portion of the residue which remains undissolved is probably resins and terpenes). To the hot solution add hydrochloric acid until litmus shows acidity, and allow to cool.

The free fatty acids will generally solidify in a cake on cooling but may remain liquid, in either case they can easily be separated by filtering through a wet filter-paper.

The acid filtrate is examined for glycerin (it will contain considerable quantities of potassium chloride) by evaporating to the smallest possible volume on a water-bath and applying the following tests for glycerin to portions of the residue.

(1)   heat with fragments of acid potassium sulphate for several minutes.

Glycerin gives a very pungent acrid characteristic smell (smell of acrolein).

(2)   Add a few drops of copper sulphate solution and then excess of potash solution.

If glycerin is present a deep blue liquid is produced instead of precipitate of cupric hydroxide.

If much glycerin is present in the residue it may be recognised by its physical characters.

[In the case of palmitin, one of the constituents of palm oil, the changes would be represented by the following equations, which may be considered typical for vegetable oils and fats.

Palmitin (glyceryl tripalmitate) $= C_3H_5(C_{16}H_{31}O_2)_3$
$$\text{Pot. palmitate (soluble in water).}$$
$$C_3H_5(C_{16}H_{31}O_2)_3 + 3KOH = 3C_{16}H_{31}O_2K + C_3H_5 (OH)_3$$
$$C_{16}H_{31}O_2K + HCl = C_{16}H_{31}O_2H + KCl.$$
$$\text{Palmitic acid (insoluble in water).}$$

If saponification does not take place easily with aqueous potash, alcoholic potash may be substituted, and the heating must be done under a reflux condenser : in this case the alcohol must be distilled off before acidifying with hydrochloric acid.]

Quantitative examination.

Determination of total oils and fats.

Proceed as in qualitative examination, but weigh the

residue obtained on evaporating off the solvent; as soon as the weight is constant, this may roughly be taken as the weight of oils and fats.

If any considerable amount of unsaponifiable residue remains after treatment with alkali, this must be washed, dried and weighed, and its weight subtracted from the total residue before calculating as total oils and fats.

Determination of free fatty acids.

The weight of these may be obtained by weighing the dried cake, or residue insoluble in water, after acidifying the products of saponification.

Determination of glycerin.

A convenient and fairly accurate process, applicable in these cases, is based on the power of glycerin to dissolve cupric hydroxide in alkaline solution.

The acid filtrate, from which free fatty acids have been removed, is neutralised with soda, and then rendered strongly alkaline by the addition of 10 c.c. of a strong soda solution; a dilute solution of copper sulphate is then run in with constant agitation until a permanent precipitate of cupric hydroxide is obtained.

Similar experiments are then made with the same quantities of water and soda to determine (by means of a standard solution of pure glycerin) how much glycerin corresponds to the solution of the amount of cupric hydroxide noticed in the original experiment, which will be the amount of glycerin in the products of saponification.

[Various modifications of this process are used, but the method described is one of the simplest, and is quite accurate enough for comparative experiments where the differences in the amount of glycerin are likely to be considerable.]

Experiments on Oils and Fats in germinating Seeds.

(1) Determine the oils and fats in dry seeds of *Lepidium sativum.*

(2) Determine the oils and fats in seedlings (dried at 100° C.) which have germinated and grown for about 15—20 days.

(3) Determine the oils and fats in seeds (dried at 100° C.) which have just commenced to germinate.

CHAPTER XII.

TANNINS AND GLUCOSIDES.

FULL particulars of the chemistry of tannins may be obtained from the following works :

WATTS. *Dictionary of Chemistry* IV.

TRIMBLE. *The Tannins.* Vol. I. Philadelphia, 1892. Vol. II. 1894.

PROCTER. *A Text-book of Tanning.* Spon, 1885.

On the Physiology of Tannins, read :—

KRAUS. *Grundlinien zu einer Physiologie des Gerbstoffes* (Leipzig, 1887).

REINITZER. Bemerkungen zur Physiologie des Gerbstoffes, *Ber. d. deut. bot. Ges.* VII. (1889).

GARDINER. Function of Tannin in Vegetable Cells. *Proc. Camb. Phil. Soc.*, 1884.

WAAGE. (Distribution, etc. of Phloroglucin.) *Ber. d. deut. bot. Ges.* VIII. (1890).

On the solubility, etc. of glucosides consult :

BEILSTEIN. *Handbuch der organischen Chemie*, 2nd ed., vol. III. ; Glykoside.

Tannins and Glucosides.

These substances will be completely extracted by alcohol of ·850 Sp. G., and will therefore be present in the alcohol extract. (Extract No. II.)

After evaporating off the alcohol, taking up with water and filtering, the solution has to be examined for tannins, glucosides, and sugars, but the whole of the sugars will rarely be present in this solution.

Some proteids may pass into this solution but they are generally rendered insoluble by the treatment, and if they are found to be present after removing the tannins they can be precipitated by alcoholic mercuric chloride (as in removing peptones) before examining for sugars : small quantities of amides if present may be neglected.

It sometimes happens that the solution is rather strongly acid, in this case it should be exactly neutralised with dilute soda before commencing examination.

Under the heading of tannins and glucosides are included a large number of different compounds which are in many respects widely different in properties, although possessing certain characters in common, and it is consequently rather difficult to give general instructions.

It is assumed, in treating of these compounds, that it is not desired to make experiments concerning their relations to metabolism ; but from any point of view it is of the greatest importance that the processes for their complete removal from solution should be thoroughly studied. Tannins and most glucosides readily give the reduction of Fehling's, Sachsse's, and the other solutions commonly used as tests for sugars; and since they are liable to split off glucoses under the action of acids, etc., it is not too much to say that it is hardly ever possible to ascertain certainly whether free glucoses were, or were not, originally present without first removing tannins. Microchemical tests for reducing sugars are useless unless it has first been shown conclusively that tannins are absent; all tannins, not only those which are glucosides, readily reduce Fehling's, etc., solutions.

It may in some cases be of interest to ascertain whether a given tannin is a glucoside or not, since there can be little doubt that the glucose which can be split off from such tannins is plastic material. This is not quite so easy as might seem apparent, but a satisfactory and fairly simple process is given on p. 273, by which this can be accomplished.

If it is required to compare the amounts of tannins in two extracts, one of the modifications of the permanganate process can be used—full details are given in Trimble, *The Tannins*, pp. 48—51, or Sutton, *Volumetric Analysis*.

It is generally sufficient to ascertain whether tannins are present, and then to determine which of the methods given is best adapted for their removal.

The addition of basic lead acetate will almost certainly carry down the whole of the tannins, many glucosides, and any proteids, etc. present, but the precipitate is very liable to contain the glucoses as well, and it is not at all easy to wash them out with alcohol or water. The use of large quantities of water for washing the precipitate is particularly to be avoided, as the precipitates of tannins with metallic salts appear to be more or less decomposed by pure water. It is better therefore not to resort to this method unless absolutely necessary.

Most of the natural tannins resemble in their properties the substance described as mimo-tannic acid (tannin from catechu and various species of *Acacia*), while they differ from commercial gallotannic acid, which gives several rather peculiar reactions and is not a good type of the general characters of vegetable tannins.

Phloroglucin is present in many cases in woody tissues (an account of its distribution is given by Waage, *loc. cit.* p. 265), but it is not probable that it is a plastic substance.

It would occur in the alcohol extract if present, and, as it reduces Fehling's, etc. solutions, should be tested for where woody tissues have been extracted.

It is easily removed if present by shaking with ether before examining for tannins and glucosides, which are not soluble in ether.

Glucosides.

After the removal of tannins and proteids (if present) portions of the solution may be tested for glucosides if it is supposed they are likely to be present.

The reactions and solubility of the various glucosides likely to occur must be studied, and if they can be removed by shaking with an immiscible solvent it is best to proceed in this way (the process should be carried out in the same way as in removing tannins with acetic ether).

If a glucoside should be present which is not removed

from aqueous solution by any immiscible solvent it must be precipitated by some appropriate reagent and the excess of reagent removed.

Before examination, the original extract must be evaporated till all alcohol is completely driven off and the residue taken up with water, and filtered if necessary. Any residue insoluble in water may be assumed to be resins, etc., and neglected.

If the solution is acid it should be exactly neutralised with dilute soda before testing.

Qualitative tests for Tannins.

Test portions of the neutral aqueous solution with :—

(1)   A few drops of 'neutral' ferric chloride (large excess must be carefully avoided).

Blue-black or dull green coloration shows tannins.

(2)   A few drops of solution of potassium ferri-cyanide and ammonia.

Reddish-brown coloration changing to brown shows tannins.

(3)   Gelatin solution.

Dirty white precipitate shows tannins.

(4)   Lime-water $(Ca(OH)_2)$.

Blue, brown, or red colour or precipitate shows tannins.

(5)   Uranium acetate.

Brown precipitate, or reddish-brown or brown colour shows tannins.

Qualitative tests for Phloroglucin.

[Where a woody tissue has been extracted, phloroglucin should be tested for in the aqueous solution obtained from the alcohol extract as described above.

Shake a portion of aqueous solution with ether, separate the ether layer and evaporate off the ether, take up the residue with water.

Test portions of the neutral aqueous solution so obtained with

(1)   Ferric chloride.

Deep violet colour shows phloroglucin.

(2)   Freshly cut pine wood and hydrochloric acid.

Reddish-violet colour shows phloroglucin.

(For further tests for Phloroglucin see BEILSTEIN, 2nd ed. vol. II.   Phloroglucin.)

If phloroglucin is present the whole of the aqueous solution may be shaken with ether: the lower layer drawn off and warmed (to expel ether) can then be used for testing for tannins, etc.]

Removal of Tannins before examining for Sugars.

As explained above all vegetable tannins do not behave in the same way to reagents, excepting perhaps to gelatin and lead acetate; it is therefore advisable to try the reactions of a small portion of the solution before deciding which method to adopt for the removal of the tannins.

The methods may be tried in the following order

till one is found which gives a filtrate quite free from tannin, as shown by ferric chloride, gelatin solution, lead acetate, etc.

(α) Shaking with acetic ether (ethyl acetate). Add about half the volume of acetic ether and shake thoroughly, allow the layers to separate and draw off the lower layer. To the separated lower layer add again about the same volume of fresh acetic ether and repeat the process.

Warm the separated lower layer (aqueous solution) on the water-bath till all smell of acetic ether has disappeared and test portions of the cold solution for tannin as above.

If the solution gives no tannin reactions the tannin is completely removed by ethyl acetate, and this method can be applied to the whole of the solution.

If on the other hand the solution still gives the tannin reactions, probably acetic ether will not remove the whole of the tannin, and it will be better to try some other method, such as (β) or (γ).

(β) Shaking with magnesia or lead carbonate.

Add freshly ignited magnesia or pure lead carbonate (about 4 grs. per 100 c.c. of liquid), shake thoroughly and allow to stand for three or four hours with frequent shaking. Filter and test the filtrate for tannins. This process seldom fails to completely remove tannins from an aqueous solution, but in some cases a considerable quantity of Pb or Mg salts soluble in water may be produced.

If all the tannins are not removed by this process it is

generally better to proceed at once to (γ) than to try precipitating with other metallic salts.

(γ) Shaking with hide powder or strips of raw hide (gelatin method).

The hide powder or strips of raw hide are thoroughly soaked in cold water, which is frequently changed, and then added to the tannin solution in considerable quantity, and allowed to remain in it for about 12—20 hrs. The whole of the tannins are absorbed by the solid mass but a considerable amount of organic matter nearly always goes into solution during the process from the hide, and before examining the filtrate for sugars it must be evaporated to dryness on a water-bath and taken up with 90 p.c. alcohol. The alcohol is then completely distilled off and the residue taken up with water.

To determine whether a tannin is a glucoside or not.

If soluble in acetic ether the tannin can be separated from free glucose by this means and the residue from evaporation of the acetic ether used in the experiment. If the tannin is not soluble in acetic ether it must be precipitated by lead acetate, or some other convenient metallic salt, and the well-washed precipitate used.

The residue or precipitate is heated with 2 p.c. hydrochloric acid for about two hours on a water-bath with a reflux condenser—allowed to cool and filtered. The solution is treated with basic lead acetate as long as it causes a precipitate, filtered and the excess of lead removed by

H_2S: after removing H_2S by warming, the solution is neutralized exactly and tested for glucose by ordinary tests (see p. 282). If the presence of much glucose is indicated, the original tannin was certainly a glucoside, but if only a small quantity of glucose is present, another experiment should be tried with a more carefully purified residue or lead precipitate.

Glucosides.

After the removal of tannins some glucosides may still remain in solution. No general method can be given for detecting these substances, but if their presence is suspected, special tests may be tried for those glucosides likely to be present in any particular case.

A glucoside can generally be separated from aqueous solution by shaking with some appropriate solvent immiscible with water, such as amyl alcohol, benzene, chloroform, etc. (not ether), and it can then be identified by evaporating off the solvent and applying tests to the residue.

E.g. To detect salicin.

The aqueous solution is shaken with an equal volume of amyl alcohol, which extracts this glucoside completely from solution in water.

The amyl alcohol is then separated from the water and a part of the solution is cautiously heated till all the solvent (amyl alcohol) is evaporated off and a dry residue obtained. Salicin may be identified in this residue by its reaction (intense red colour) with strong sulphuric acid, and also by dissolving another part of this residue in water and applying the 'Saliretin' test to the aqueous solution[1].

Sugars, etc.

After removal of tannins (also of glucosides and proteids if necessary), the solution may be examined for sugars.

The detection and estimation of sugars in this solution is carried out in exactly the same way as described for extract III. (cold water) after removal of dextrins (see p. 281) with which it may be mixed for examination.

The whole of the sugars may occur in this solution.

Experiments.

(1) Make alcohol (·850) extract of willow bark (*Salix viminalis*) after previously extracting with benzene to remove resins, colouring matter, etc.

Test the extract for

(*a*) Tannins.   (*b*) Glucosides (Salicin).   (*c*) Sugars.

(2) Compare the amounts of sugars which can be extracted by ·850 alcohol in young and in ripe fruits of *Musa sapientum*, L.

(3) Show that the tannin extracted by ·850 alcohol is a glucoside.

Make alcoholic extracts of material dried in each case at 100° C.—and compare the amounts of sugar in solutions after complete removal of the tannins.

It is not necessary in this case to previously extract the material with benzene or petroleum ether.

---

[1] See Fresenius, *Qualitative Analysis* 10th ed. p. 436.

CHAPTER XIII.

DEXTRINS AND SUGARS, GLUCOSES, CANE-SUGAR, MALTOSE, ETC.

BEFORE commencing work in this section full information as to the properties and characteristic reactions of the commoner carbohydrates should be obtained.

For this purpose consult :—

TOLLENS. *Handbuch der Kohlenhydrate.* Breslau. 1888.

BEILSTEIN. *Handbuch der organischen Chemie.* 3rd Ed. Hamburg und Leipzig, 1893. (Vol. I. Kohlenhydrate.)

WATTS. *Dictionary of Chemistry* (ed. Muir and Morley). Articles, Dextrins, Vol. II.; Sugars, Starch, Vol. IV.

Read papers by :—

BROWN and MORRIS. *J.C.S.* trans. LVII. (1890) 458, and *J.C.S.* trans. LXIII. (1893) 604. (Abstract in *Annals of Bot.* VII. No. 2.)

E. SCHULZE. *Landw. Vers.-Stat.* XXXIV. (1887.)

SCHIMPER. *Bot. Zeit.* (1885.)

PRUNET. *Compt. rend.* CXV. (1892.)

KULISCH. (Sugars in Fruits.) *Landw. Jahrb.* XXI. (1892).

Soluble Carbohydrates (Dextrins, Sugars, etc.).

These will partly be present in the cold-water extract but some of the sugars will be found in the previous alcohol extract as already pointed out.

The examination of the solutions will depend on the nature of the results required; the simpler methods will frequently be sufficient for problems on metabolism, but methods for dealing with cases where more detailed information is required are also given.

Since the common sugars, glucoses, cane-sugar, and maltose are all undoubtedly plastic substances, it is often sufficient to determine these together as fermentable sugars and to calculate the result as 'glucoses.' The simple fermentation processes can often be used, as it is but rarely that solutions are obtained (after removal of tannins, etc., as above described) which contain sugars that do not ferment on the addition of yeast.

Dextrins do not ferment, or hinder fermentation, and it is not therefore necessary to separate them from a solution in using the fermentation methods.

Dextrins interfere much with determinations of sugars by the volumetric 'reducing' processes (in many cases they act as reducing agents), and although the precipitation by alcohol is tedious and troublesome it is generally better to use it, and to make the estimations with solutions free from dextrins.

A full discussion of the methods for estimating sugars in mixtures of glucoses, cane-sugar, and maltose need not be entered into here, but it may be stated that the process given on p. 285 gives satisfactory results, and although many alternative methods may be used to confirm the results it will in most cases be quite sufficient.

It is desirable that experience should be gained in the use of one method before attempting others, but when results can be confirmed by the application of an independent method, it may sometimes be satisfactory to use more than one for the same object.

If any proteids, etc., are found in the solution after precipitation of the dextrins, they should be removed by mercuric chloride as in previous cases, but this will rarely occur.

Qualitative test for fermentable Sugars.

To a portion of the solution add a small quantity of active yeast suspended in pure water and allow it to stand in a warm place for several hours in a flask fitted with a bent tube, the free end of which passes into a solution of baryta water protected from access of air.

If fermentable sugars are present an abundant precipitate will be caused in the baryta-water (from CO_2 evolved), and alcohol can be detected in the contents of the flask by distilling off a small portion and testing the distillate by iodoform or other reactions.

Quantitative.

To estimate the amount of fermentable sugars it is usual to absorb the CO_2 evolved in some appropriate form of apparatus which can be weighed before and after absorption of the gas; or to estimate the CO_2 by loss of weight of an apparatus from which the dry gas only can escape.

The problem is the same as that of the estimation of CO_2 in carbonates, and full directions for the process are given in Fresenius, *Quantitative Analysis*, 7th ed. (Carbonic Acid).

It is necessary to make a parallel experiment with the same quantity of yeast in pure water and to subtract the amount of CO_2 obtained in this experiment from the total; the yeast itself always evolves some gas on standing for so long a time as is generally necessary for complete fermentation.

A convenient way of performing the estimation is to use two nitrometers, introducing the solution and yeast into one (over mercury) and the same quantity of yeast in water into the other (over mercury), at the end of the experiment the volumes of gas are read off and the weight deduced from the volume of gas obtained.

Where the amount of sugars is small it is better to estimate the alcohol produced rather than the CO_2, and this may be done in all cases as a control experiment. If the amount of alcohol in the distillate is considerable it is determined from specific gravity tables, but if small by Dupré's method of oxidising to acetic acid and estimating the acetic acid by standard alkali.

(For Dupré's process see *J. C. S.*, vol. XX.)

The following numbers may be used in calculating weights of glucose in fermentation experiments.

$$CO_2 \times 2\cdot16 = \text{glucose.}$$
$$\text{Alcohol} \times 2\cdot07 = \text{glucose.}$$
$$\text{Acetic acid} \times 1\cdot58 = \text{glucose.}$$

[Allowance is made for a deficit (products other than CO_2 and alcohol) of 5·5 p.c.]

The sugars in solution may also be estimated by finding the loss of specific gravity of the solution on fermentation (after distilling off alcohol).

The specific gravity of the solution is taken by an accurate hydrometer or with a specific gravity bottle, the volume exactly noted, and yeast added. After the fermentation is complete the whole (or a definite part) is filtered and the alcohol distilled off, after which the liquid is made up exactly to its original volume with distilled water, and the specific gravity again taken.

The loss of specific gravity is due to the destruction of the sugars, and the amount of the latter can be calculated as shown below.

The specific gravity of dilute sugar solutions increases by about ·00386 for every gram of sugars in 100 c.c. of solution (at 15° C.), if therefore the specific gravity of the solution had diminished by ·008 after fermentation we should conclude that the sugars in 100 c.c. of the original solution $= \dfrac{·008}{·00386}$ grams.

Or the percentage by weight of sugars may be obtained by the use of tables[1] (of which a number have been published) shewing the relation between specific gravity and percentage by weight of sugars.

At a temperature of 18°—20° C. fermentation will usually be complete in thirty to forty hours, but it is safer to let the action continue for three days.

[1] See *Chemiker-Kalender* (1893, pp. 78—80).

The appearance of the yeast (which forms a layer at the bottom of the liquid when fermentation is complete) and absence of frothing on the surface will easily show when the change is completed.

Further examination of aqueous extract (pentoses), dextrins, (inulin), glucoses, cane-sugar, maltose, (mannite).

Qualitative.

Concentrate a portion of the aqueous solution to a small volume—a few c.c.—on the water-bath, add a large excess of strong alcohol and allow it to stand till the precipitate settles: if the addition of more alcohol causes a further precipitate allow it to settle again. The alcoholic solution can, in most cases, be completely decanted from the precipitate. If filtration is necessary, asbestos should be used.

[The precipitate is generally dextrins and need not be further examined, except for inulin, which would be precipitated with the dextrins. To test for inulin dissolve a portion of the precipitate in water, add a few drops of strong hydrochloric acid, boil, cool, and add a few drops of alcoholic solution of phloroglucin. Yellow-brown colour indicates inulin[1].

If this reaction gives a positive result, dissolve the rest of the precipitate in water and add baryta-water as long as it causes a precipitate : collect and wash the precipitate, suspend it in water and decompose with a current

[1] Green, *Annals of Bot.*, Vol. I. p. 233.

of CO_2: filter, and evaporate down the solution : crystals should be obtained if inulin was present.]

The alcoholic filtrate is evaporated till the alcohol is completely removed and the residue dissolved in water. The alcohol may be condensed and preserved.

The solution thus obtained is tested :—

    (a)   for reducing sugars, glucoses, maltose,

    (b)   for cane-sugar,

    [(c) for mannite and pentoses.]

(a) Test portions with

    (1)   Fehling's solution

    (2)   Sachsse's   ,,

    (3)   Barfoed's   ,,

  (4 p.c. crystallized cupric acetate + 1 p.c. acetic acid.)

    (4)   Phenyl-hydrazin reaction.

If positive results are obtained with (1) and (2) but not with (3) and not easily with (4) it would indicate that only maltose is present, but this is rarely the case.

To ascertain whether maltose is present together with glucoses, it is necessary to make roughly quantitative experiments, and it is convenient to arrange these so as also to include the testing for cane-sugar.

Divide the solution into several equal parts.

(1) Use one part to determine the original 'reducing power' of the solution.

(2) Heat another part with citric acid and ascertain whether the 'reducing power' of the solution has been increased or not by heating with this acid.

(3) Heat another part with dilute hydrochloric acid and ascertain whether the 'reducing power' has been further increased or not, i.e. if the value obtained for the 'reducing power' is greater than that obtained in (2).

Since cane-sugar only is inverted by citric acid and both cane-sugar and maltose by hydrochloric acid, it follows that an increase in (2) indicates the presence of cane-sugar, and a further increase in (3) that of maltose.

The exact details for performing these experiments are given in the next section under quantitative examination—for qualitative evidence rough estimations will suffice.

If the presence of cane-sugar and maltose is indicated as above, attempts should be made to obtain crystals of the cane-sugar and an osazone of the maltose.

To obtain crystals of the cane-sugar add, to a portion of the solution, strontia-water $Sr(OH)_2$ in considerable quantity and filter; evaporate down the filtrate till a precipitate (yellowish amorphous masses) begins to separate out, and let it stand for some time. Collect the precipitate, suspend in dilute alcohol and decompose with a current of CO_2; filter, concentrate the filtrate if necessary, and add strong alcohol till it just begins to cause turbidity. Stir in a small crystal of solid cane-sugar and allow it to stand.

To obtain maltosazone from maltose in presence of glucose, evaporate the solution to small bulk and add phenyl-hydrazin and sodium acetate as in testing for glucose. Heat on the water-bath for one hour with reflux condenser, and allow the products to stand in the cold for several hours. Filter off the glucosazone (which is nearly insoluble in cold water) and evaporate the solution to a small bulk on the water-bath and allow it to cool; maltosazone (which is soluble in water) should separate out if maltose was present in the original solution.

[The maltosazone obtained in this way will not be pure, for satisfactory identification it should be recrystallised from alcohol and the melting point determined.]

This reaction in connection with the evidence obtained by an increase of reducing power on complete inversion, as compared with the reducing power after inversion with citric acid, will show satisfactorily whether maltose is present or not.

(c) Mannite and Pentoses.

Mannite.

A portion of solution is completely fermented with yeast, filtered, evaporated to dryness, and the residue is taken up with absolute alcohol. Addition of ether causes a precipitate if mannite is present.

If ether causes a precipitate, the precipitate should be collected and recrystallised from alcohol. Mannite crystallises well and may be recognised by the appearance of the crystals.

[Dulcite gives the same reactions, but can be easily distinguished from mannite by giving mucic acid when oxidised with nitric acid. Dulcite is of rare occurrence.]

Pentoses.

A portion of aqueous solution is distilled with hydrochloric acid, and the distillate tested for furfurol (furfur-aldehyde) by the colour reaction with aniline acetate.

[BROWN and MORRIS did not find pentoses in leaves of *Tropæolum*: but compare the paper by DE CHALMOT, *Amer. Chem. J.* xv. (1893).]

Quantitative.

It is not generally necessary to estimate dextrins, but if necessary the precipitate by alcohol may be estimated by the method described for starch; or it may be dissolved in water, dried at 110° C., and weighed as anhydrous dextrins.

Estimation of glucoses, maltose, cane-sugar.

For an account of the volumetric methods of estimating the reducing power of solutions by Fehling's, Pavy-Fehling's, or Sachsse's solutions, consult SUTTON, *Volumetric Analysis*, pp. 259—269, where full details are given for preparing the standard solutions, and performing the operations.

[For standardising these solutions, pure anhydrous glucose (dextrose) can now be obtained from dealers in pure chemical reagents.]

Divide the solution in which the sugars are to be determined into three equal parts; 100 c.c. each is generally convenient.

(1)　Use one part for the determination of the original 'reducing power' of the solution.

(2)　Use another part for the determination of the 'reducing power' after 'inversion' with citric acid.

(3)　Use another part for the determination of the

'reducing power' after complete 'inversion,' i.e. inversion with 2 p.c. hydrochloric acid.

To invert with citric acid add 5 grs. of solid crystallised acid for every 100 c.c. of the solution and heat on the water-bath for one hour under a reflux condenser; then exactly neutralise and if necessary make up to the original or some definite volume.

For complete inversion add to the solution sufficient hydrochloric acid to make 2 p.c. of acid and heat on the water-bath for three hours under a reflux condenser; then exactly neutralise, etc. as after inversion with citric acid.

The determinations of 'reducing power' in (1) (2) and (3) should be made as accurately as possible.

[It is convenient to state results in 'grs. glucose per 100 c.c. of solution,' meaning that 100 c.c. of solution has the same reducing power as would be possessed by a solution containing so many grs. glucose.]

The calculation is easy when the method is thoroughly understood.

The reducing power is stated in grs. glucose in each case per 100 c.c. of solution.

Original reducing power $= a$.

Reducing power after inversion with citric acid $= b$.

Reducing power after inversion with hydrochloric acid $= c$.

$b - a =$ glucose from inverted cane-sugar.

$\therefore (b - a) \times \cdot 95 =$ cane-sugar.

$c - b =$ glucose corresponding to increased reducing power of maltose inverted.

$\therefore (c - b) \times 2\cdot 32 =$ maltose.

$a -$ (maltose $\times$ ·62) = glucose.

[The numbers used above are obtained from these data :
342 grs. $C_{12}H_{22}O_{11}$ give 360 grs. $C_6H_{12}O_6$ on inversion.

$\therefore$ glucose $\times$ ·95 = cane-sugar or maltose.

1 gr. maltose has same reducing power as ·62 grs. glucose.

1 gr. maltose gives 1·05 grs. glucose on inversion.

$\therefore$ 1 gr. maltose on inversion gives increased reducing power = ·43 grs. glucose (1·05—·62).

$$\frac{1}{·43} = 2·32.$$

$\therefore$ increase of glucose by inversion of maltose
$$\times\ 2·32 = \text{maltose}\ ]$$

An actual example will illustrate use of the above.
100 c.c. of an extract :—

(a)   reduced 30 c.c. of Fehling.

(b)   after inversion with citric acid and neutralising, 38 c.c. Fehling.

(c)   after inversion with hydrochloric acid and neutralising, 45 c.c. Fehling.

(10 c.c. Fehling = ·05 grs. glucose.)

Then $b - a = 38 - 30 = 8$ c.c. Fehling = ·04 grs. glucose.

$\therefore$ *cane-sugar* = ·04 $\times$ ·95 = ·038 grs.

$c - b = 45 - 38 = 7$ c.c. Fehling = ·035 grs. glucose.

$\therefore$ *maltose* = ·035 $\times$ 2·32 = ·081 grs.

$a = 30$ c.c. Fehling = ·150 grs. glucose.

$\therefore$ *original glucose* = ·150 grs. $-$ (·081 $\times$ ·62)
$$= ·1\ \text{gr.}$$

Therefore 100 c.c. of extract contain,

Glucose ·100 gr.

Cane-sugar ·038 gr.

Maltose ·081 gr.

Experiments on Sugars.

Qualitative.

I.   Examine leaves of *Tropæolum majus* for glucoses, cane-sugar, maltose.

Extract the dry material (1) with benzene, (2) with alcohol, (3) with cold water.

After removal of tannins, etc. from extract (2) mix the solution with extract (3) and examine as directed on p. 282.

II.   Examine (*a*) leaves of actively growing *Beta vulgaris* which have been killed with chloroform and dried.

(*b*)   roots of do. which have been stored at the end of summer.

Examine as directed for the leaves of *Tropæolum*.

Quantitative.

I.   Compare the amounts of fermentable sugars (calculated as glucoses) in the materials used for Qualitative, II.

Extract at once with cold water and use the methods described on p. 278 after removal of tannins by shaking with lead carbonate (see p. 272).

II.   Compare amounts of sugars in (*a*) leaves of *Beta vulgaris* which have been killed by chloroform whilst in a state of active assimilation.

(*b*)   leaves of do. which have been cut off from the plant during active assimilation, kept in the dark for forty-eight hours, and then killed with chloroform.

Extract with (1) benzene, (2) alcohol, (3) cold water.

After removing tannins, etc. from (2) mix the solution with extract (3) and estimate glucoses, cane-sugar, and maltose by the process described on p. 285.

CHAPTER XIV.

Starch and Cellulose.

Read:—

Wohl. *Ber. d. d. chem. Ges.* XXIII. (1890).

E. Schulze. *Ber. d. d. chem. Ges.* XXIV. (1891).

Winterstein. *Zeit. physiol. Chem.* XVII. (1892).

Scheibler and Mittelmeier. *Ber. d. d. chem. Ges.*
XXIII. 2 (1890); *ibid.* XXIV. 1 (1891); *ibid.* XXVI. 3 (1893).

Estimation of Starch.

A discussion of methods suitable for estimation of
starch in leaves and plant tissues will be found in Brown
and Morris' paper, *J. C. S.* LX. (1893), pp. 622—633.

Fair results may sometimes be obtained by using modifications of Sachs' iodine method[1] colorimetrically, but
it is frequently necessary to resort to 'chemical methods.'
These are based on the insolubility of starch in cold water
and its solubility in dilute acid, or in an extract containing
diastase, cellulose being insoluble under the influence of
these reagents.

[1] See Part I, p. 21.

Hot dilute acid (1 p.c.) is frequently used, being much more convenient than diastase, but there is reason to believe that this method under some circumstances yields results seriously too high, as the acid is not without influence on the cellulose (compare E. Schulze, *Ber. d. d. Chem. Ges.* XXIV. (1891), p. 2277, but also Winterstein, *loc. cit.*).

With both reagents the starch is first converted into soluble products (dextrin and maltose, or dextrin, maltose and glucose), and the products are then further hydrolysed to glucose by continued action of stronger acid. The latter stage of the process is liable to error in both directions, the hydrolysis may not be complete or the glucose may be partly destroyed by 'reversion.' To obtain good results it is necessary therefore to adhere strictly to the instructions given for any process.

Heating for half-an-hour with 1 p.c. sulphuric acid in a sealed tube at 120° C. (pressure about 2 atmospheres) gives good results with pure starch.

A full account of Allihn's and other methods for estimating starch is given in Tollens' work.

Starch.

The whole of the soluble products obtained by limited hydrolysis of starch will be contained in the dilute acid extract (Extract IV.), and there will generally be little else in this extract. Any proteids etc., which it may contain may be neglected.

Qualitative.

The original material is tested for solid starch either microchemically or by Sachs' iodine method.

Quantitative.

A measured quantity of acid solution (containing 1 p.c. sulphuric acid) is placed in a strong glass tube and the open end carefully sealed. The tube is then heated and kept at 120° C. for half-an-hour. When cool the tube is carefully opened, the contents washed out and made up to a known volume with distilled water, and the glucose estimated, after neutralising, by one of the volumetric processes.

Glucose found × ·9 = starch (anhydrous), ($C_6H_{10}O_5$).

If diastase (malt-extract) instead of dilute acid is used to extract the starch from the material, the solution obtained (containing dextrins and maltose) may be treated in the same way to convert all the first products of the hydrolysis of the starch into glucose.

In a measured quantity of the solution, to which sufficient sulphuric acid has been added to make 1 p.c. acid, the glucose is estimated after heating in a sealed tube at 120° C. for half-an-hour.

When this method is used a correction has to be made for the reducing power of the substances produced by the action of the acid at 120° C. on the malt extract.

For this purpose a quantity of the malt extract equal to that taken for dissolving out the starch from the original material must be treated as directed above, and the glucose produced in the operation determined.

The amount of glucose found in this last experiment must be subtracted from the total amount found in the actual experiment, as the difference between these amounts represents the glucose produced from the starch in the original material.

In the few cases in which it is necessary to estimate cellulose the residue from starch estimation can be used. The residue is shaken with ammonia, washed and then treated with dilute bromine water as long as the latter is decolorised, if much bromine is required it is better to treat again with ammonia and repeat the oxidation with bromine, the residue is then washed with ammonia and finally with water, dried at 100° and weighed as cellulose.

The *rationale* of the process is that substances other than cellulose are oxidised by the bromine water to products soluble in alkali, but the cellulose is not altered.

Experiments on Starch.

Quantitative.

I.   Estimate the amount of starch in the potato by the processes given on p. 292.

Dry at 100° C. and :—

(*a*)   Extract (1) with alcohol, (2) with cold water, (3) with 1 p.c. sulphuric acid at 100° C.

(*b*)   Extract (1) with alcohol, (2) with cold water, (3) with malt extract (diastase) at 50—55° C.

(*c*)   Also estimate the starch in another portion of the same by specific gravity and tables.

This can be conveniently performed by floating the potatoes in a saturated salt solution and adding water till they just sink; the specific gravity of the fluid, taken with an accurate hydrometer, then equals the sp. gr. of the potatoes.

Tables showing the p.c. by weight of dry starch in potatoes, corresponding to different specific gravities, have been constructed[1].

This method only applies to potato tubers.

II. Compare the amount of starch in leaves of *Acer pseudoplatanus* which have been collected on a warm evening after a bright day (and dried after killing with chloroform) with the amount in leaves of the same species collected early next morning and treated in same way.

Use the method given in I. (*a*).

III. Determine the starch in grains of wheat: and in grains of wheat which have germinated and been kept in the dark for three days before killing.

Use the method given in I. (*a*).

Let the grains germinate on damp sand, and after keeping three days in the dark take whole plants (including the residue of the grains, wash thoroughly, and dry at 100° C.

[1] See *Chemiker-Kalender* (1893, p. 377).

CHAPTER XV.

ORGANIC ACIDS AND SALTS.

Consult :—

FRESENIUS. *Qualitative Analysis.* 'Organic acids.'

Read :—

DE VRIES. *Bot. Zeitung* 1879.

KOHL. *Bot. Centralblatt* XXXVIII. 1889.

WEHMER. *Bot. Zeitung* 1889.

   ,,   ,,   ,,   1891.

   ,,   *Landw. Vers.-Stat.* XL. (1892).

DOEBNER. *Ber. d. d. chem. Ges.* (XXVII.) 1894.

Organic acids.

It is doubtful whether any of these compounds are plastic, and it is almost certain that the lower members of the series such as acetic, malic, oxalic, etc. are not, but many of them are certainly indirectly useful in metabolism.

It is not generally necessary to ascertain exactly to what acids the acidity in any case is due, but it may be required to compare the amounts of acid present in tissues, in the free state or as acid salts. If it is required also to determine the combined acids in neutral salts, this may be done by precipitating with lead acetate, suspending the precipitate in water and decomposing with H_2S; the free acid can be estimated in the solution (after filtering and warming to expel H_2S) by titration with standard baryta-water, using phenolphthaleïn as the indicator.

Acids of which the calcium salts are insoluble in water will not be included in this estimation of the total acid, but if present they can be determined in the same way as for calcium oxalate.

Qualitative examination for Organic Acids.

For this purpose it is better to make a special extract of fresh tissues, by pressing out the juice as completely as possible, washing the residue with cold water, and diluting up to a definite volume.

Organic acids may occur free, or as acid salts, or as neutral salts. Ethereal salts (esters) of organic acids may also occur, but as these are insoluble in water and soluble in petroleum ether, benzene, etc., they are extracted and examined with oils and fats.

The identification of organic acids in a mixture is rather a difficult problem: full instructions will be found in Fresenius, *Qualitative Analysis*.

It must be remembered that salts of sulphuric, phosphoric and hydrochloric acids are nearly always present.

Students who have not had some practical experience in testing for organic acids in mixtures, should practise the methods on mixtures of known composition before attempting to identify particular acids in actual extracts.

To determine the 'acidity' of an extract.

Any exactly standardised caustic alkali may be used with phenolphthaleïn for indicator. $\frac{N}{10}$ baryta is generally convenient and sometimes gives a sharper reaction than soda or potash. The results may be calculated in grs. BaO neutralised, or as oxalic or acetic acids.

If an extract should be alkaline and it is desired to determine the alkalinity, it is better not to titrate directly with the standard acid, unless methyl orange is used as indicator, because the alkalinity will be partly at least due to alkaline carbonates.

The alkalinity is best determined by adding a known volume of standard acid (excess), warming for some time to expel CO_2 and then finding the excess of acid by $\frac{N}{10}$ baryta and phenolphthaleïn in cold solution.

The results may be calculated in grs. acid neutralised.

Inorganic salts.

Consult :—

FRESENIUS. *Quantitative Analysis.* 6th English edition. Special part, pp. 678—690.

FRESENIUS. *Qualitative Analysis.* 10th English edition. Ash of plants.

Read :—

PALLADIN. *Ber. d. d. bot. Ges.* IX. (1891).

LAWES and GILBERT. *J. C. S. Trans.* XLV. (1874).

HELLRIEGEL.   *Landw. Vers.-stat.* IV. (1862).
STOOD.   *Landw. Vers.-stat.* XXXVI. (1889).
KRAUS (abstr.).   *Bot. Centralblatt* XLIX. (1892).
LESAGE.   *Compt. rend.* CXII. (1891).
LOEW.   *Biol. Centralblatt* XI. (1891).
    "   *Flora.*   1892.
SCHIMPER.   *Flora.*   1890.

Inorganic salts.   Ash of tissues.

The relations of various inorganic salts to metabolism
are studied by culture experiments or by examining the
ash of various tissues. The method of culture experiments
is described in Part I., p. 58.

The best processes for obtaining the ash of a tissue are
fully discussed in Fresenius, *Quantitative Analysis*, Special
part, sixth English edition, pp. 678—690, and full details
for determining the constituents are given.

Studies on the ash of tissues are only useful for
determining the distribution of bases and of chlorine and
silica; the other acid radicles commonly present, viz. CO_2,
SO_3, P_2O_5, do not represent the proportions in which they
were present as such in the original tissue, indeed in
extreme cases they may have been entirely formed during
incineration; but estimations of P_2O_5 may sometimes be
useful, since if much P_2O_5 is found, the greater part of it
was probably present as such originally.

In the case of readily fusible ash the practice commonly
followed in the preparation of ash of sugars is very useful,
viz. sulphating, but the chlorine will be expelled and cannot
be estimated in a sulphated ash.

The addition of sulphuric acid greatly facilitates the burning of the last portions of carbon, and the process may generally be employed, but it must be remembered in this case that the ash will weigh rather more than an original ash, as CO_2 and Cl_2 are replaced by SO_3.

Qualitative examination.

Since the constituents of ash are almost invariably the same it is seldom necessary to make a qualitative examination.

The bases are potassa, soda, lime, magnesia, ferric oxide, and small quantities of alumina and oxides of manganese.

The acid radicles are carbonic, sulphuric, phosphoric, silicic acids, and chlorine.

Quantitative examination.

If a complete analysis is required the scheme given in Fresenius (see above) may be followed, and it is easy to estimate any of the constituents by modifications of this process.

Determinations of chlorine, phosphoric acid, and alkalies may easily be made.

Chlorine.

A weighed quantity of the ash is boiled with water and filtered, the chlorine can be estimated in a portion of the filtrate after exactly neutralising with nitric acid by standard solution of silver nitrate, using potassium chromate as indicator.

See Sutton, *Volumetric Analysis*, p. 112.

Phosphoric acid.

A weighed quantity of the ash is dissolved in dilute nitric acid and filtered.

The phosphoric acid in the filtrate is precipitated with ammonium molybdate. The precipitate is collected, washed and dissolved in ammonia ; the phosphoric acid in the ammonia solution may be estimated volumetrically by a standard solution of uranium acetate, after precipitating with magnesia mixture and dissolving the precipitate in dilute acid.

See Sutton, *Volumetric Analysis*, pp. 238—247.

Alkalies.

A weighed quantity of the ash is boiled with hydrochloric acid and filtered. The filtrate is evaporated till nearly all free acid is removed, and baryta water added in slight excess; filter, after heating for some time on the water-bath : the filtrate will only contain the alkalies and barium salts as the other constituents are precipitated.

Add sufficient ammonia and ammonium carbonate to precipitate all the barium, let the precipitate subside and filter: evaporate the filtrate to dryness in a platinum dish and ignite till all ammonium salts are expelled.

Weigh the residue (chlorides of sodium and potassium). If it is desired to ascertain the amounts of sodium and potassium the residue can be dissolved in water after weighing, and the chlorine determined by standard solution of silver nitrate. The weight of the two chlorides and of the chlorine being determined the proportions in which Na and K are present can be calculated.

(If $x = $ NaCl and $y = $ KCl,

$$\begin{cases} x + y = a \text{ (weight of residue)}, \\ \dfrac{Cl}{NaCl}\, x + \dfrac{Cl}{KCl}\, y = b \text{ (weight of chlorine in } a) : \end{cases}$$

from these equations the values of x and y can be found.)

Calcium oxalate.

Calcium oxalate can readily be estimated in plant tissues by extracting a weighed portion of tissue with dilute HCl, evaporating the extract to dryness and strongly igniting the residue. The ignited residue is dissolved in HCl and the calcium estimated volumetrically by the usual process, with oxalate and standard potassium permanganate.

The calcium found is calculated to calcium oxalate (CaC_2O_4 or $CaC_2O_4 + H_2O$).

See Sutton, *Volumetric Analysis*, p. 133.

Experiments on organic acids.

(1) Compare the acidity of juice pressed out of equal weights of petioles of young and old leaves of rhubarb (*Rheum rhaponticum*).

(2) Compare the acidity and amounts of sugars in juice from ripe and unripe apples.

Determine the acidity of the juices with $\dfrac{N}{20}$ baryta and phenolphthaleïn as directed on p. 297.

To determine the sugars take 50 c.c. of filtered and diluted juice, add lead acetate as long as it causes a precipitate, filter and remove excess of lead by H_2S etc. and determine the sugars as directed on p. 285.

Experiments on inorganic salts.

(1) Compare the amounts of ash which are obtained from equal dry weights of leaves of normal and etiolated potato plants.

Weigh the sulphated ash.

(2) Compare p.c. of P_2O_5 and alkalis (soda and potassa) in the ash of young leaves and of grains of wheat or barley.

Obtain normal ash and estimate P_2O_5 and alkalis as directed on p. 300.

(3) Compare the amounts of calcium oxalate in young and old leaves of *Sempervivum tectorum*.

CHAPTER XVI.

Unorganised Ferments (Enzymes).

Read :—

GREEN. *Annals of Bot.* Vol. VII., pp. 83—137.

Diastase.

LINTNER and DÜLL. *Ber. d. d. chem. Ges.* XXVI. (1893).

BROWN and MORRIS. *Loc. cit.* pp. 633—661.

(References to other papers on diastase are included by the authors of the above.)

Invertase.

O'SULLIVAN and TOMPSON. *J. C. S. Trans.* LVIII. (1890.)

Glycase.

LINTNER. *Zeit. für das ges. Brauwesen* XV. (1892.)

Unorganised ferments are soluble in water and precipitated by alcohol which does not destroy their power, but the latter is destroyed by heating above 80° C. An unorganised ferment should therefore be extracted by cold water, but this is not always possible, even when the material has been killed and is completely disintegrated, so that cases may occur where material suspended in water will show ferment action, when an extract will not.

Dilute glycerin may be used instead of water for extracting ferments; if this is used the solution is precipitated by alcohol and the precipitate dissolved in water.

The amount of ferment present in a solution cannot be determined, but the amount of a particular change produced by different solutions containing a ferment may be compared.

Qualitative examination.

The presence of ferments is shown by allowing the extract to stand in contact with the substances on which it is supposed to be capable of acting for some hours at 30°—50° C., and then testing for characteristic products.

A parallel experiment is made at the same time with another portion of the extract which has been boiled before using, to destroy the ferment; if none of the characteristic product is produced in this case, the action of the original extract may be attributed to ferment.

Quantitative.

Equal quantities of two extracts are allowed to act, under similar conditions, for equal times on equal weights of the substances to be acted on, and one of the products of the change is estimated.

The activity of the two solutions will be roughly proportional to the amounts of the product formed.

Kjeldahl has shown that methods based on measuring

the time required for transforming a given weight of starch are not reliable, since there does not appear to be a simple relation between the rate of change and amount of diastase causing it.

Or

The volumes of extracts required to completely change a given weight of the substance in a given time may be found by trial. The activity of the extracts will be roughly proportional to the volumes used.

[A discussion of processes for comparing the diastatic activity of tissues is given by BROWN and MORRIS, *loc. cit.* p. 637 ; the same principles will apply to other ferments.]

Preparation of a Ferment.   (Diastase.)

Make an aqueous extract of malt, using about 100 c.c. of water for each 10 grs. of malt taken.

Mash up the solid as completely as possible in a mortar with a little warm water, transfer to a bottle, adding more water, and shake for two hours. Then heat for half-an-hour at 50°—55° C., filter and concentrate the aqueous solution under reduced pressure to a small bulk.

To the concentrated aqueous solution add 90 p.c. alcohol so long as it causes a flocculent precipitate, ceasing the addition of alcohol when it begins to render the liquid distinctly turbid or markedly opalescent; filter off the precipitate, wash it with absolute alcohol and dry in an exhausted desiccator over sulphuric acid.

The solid is impure diastase.

Experiments on Diastatic Ferments.

(1)   Prepare a specimen of diastase.

(2) Compare the diastatic power of malt and un-germinated barley.

(3) Shew that cold water extract of malt has diastatic power, but that cold water extract of leaves of barley when filtered clear has little or no diastatic power.

(4) Compare the diastatic power of the leaves of *Pisum sativum* and *Trifolium pratense*.

Experiments on Invertase and Glycase.

(1) Shew that water in which living yeast cells have been suspended has no invertive action on cane-sugar when filtered quite free from yeast cells, but that water in which yeast cells have been killed and crushed has inver-tive power even when filtered quite clear.

(2) Shew that the addition of finely ground maize suspended in water transforms maltose into dextrose.

Glucoside Ferments.

Shew that emulsion of bitter almonds will produce saligenol from salicin.

Diastase.

(1) Prepare a specimen of solid diastase by the method given on p. 305.

Shew that the specimen is soluble in water and that it will produce dextrins and maltose from starch.

(2) Make aqueous extracts of malt and barley as described for preparation of diastase.

Compare the diastatic power of the extracts by their action on soluble starch.

[Starch solution for these experiments should be very carefully made. Rub ·2 grs. of pure dry starch into a thin cream with a little cold water and pour it into 200 c.c. of actually boiling water; continue boiling for two minutes and allow it to cool. The solution should always be freshly prepared immediately before using.]

Put 10 c.c. of the above starch solution into each of five test-tubes marked A, B, C, D, E.

To A add 1 c.c. extract + 4 c.c. water.

B	„	2 c.c.	„	+ 3 c.c.	„
C	„	3 c.c.	„	+ 2 c.c.	„
D	„	4 c.c.	„	+ 1 c.c.	„
E	„	5 c.c.	„	+ 0	„

Place the test-tubes in a water bath at 50°—55° C., and allow to stand for one hour. Take out the test-tubes and cool thoroughly, then test a few drops from each (about ·2—·3 c.c.) with a drop of iodine solution and note which tube shews the reaction to be just complete (i.e. the first which gives no blue or violet colour with iodine).

If the reaction is not complete in any of the tubes, more of the extract can be added in the same proportions as at first and the heating continued for another hour at 50°—55° C., if the same quantity was taken from each for examination.

If the reaction is complete in all the tubes more starch solution can be added in the same way. A rough preliminary experiment will easily shew about what quantities of starch solution and extract will be convenient.

Repeat the determinations with the second extracts in an exactly similar manner.

The diastatic power of the extracts will be proportional to the numbers of c.c. of each which will 'convert' a given weight of soluble starch in a given time under similar conditions : e.g. if in one set of experiments 'conversion' is just complete in B (2 c.c. extract), and in another in D (4 c.c. extract), the diastatic power of the first extract is twice that of the second.

Another method is to allow equal quantities of the extracts to act upon excess of solution of soluble starch for one hour at $50°$—$55°$ C., and then estimate the 'reducing sugars' produced.

At the end of one hour the mixtures are concentrated on a water bath, after first rapidly heating to $100°$ C. to stop further action, and the starch and dextrins precipitated together by excess of alcohol.

The reducing sugars are estimated in the filtrate, after distilling off the alcohol and taking up the residue with water, by Fehling's (or other suitable) solution (see p. 285).

The reducing power of the extracts used must be determined in each case and subtracted from the total reducing sugars found.

The diastatic power of the extracts will be proportional to the amounts of reducing sugars formed by their action on soluble starch.

Invertase.

(1)   Filter off some of the water in which active yeast is suspended, by adding some 'paper pulp' and filtering through asbestos : [it is not very easy to completely remove yeast by ordinary filtration].

The solution does *not* invert cane-sugar.

(2) Mash up some yeast thoroughly with a little water and ether (to kill the yeast cells), filter as in (1). The solution contains invertase and *will* invert cane-sugar rapidly.

Use a 2 p.c. solution of cane-sugar for these experiments and treat it for one hour at 20°—30° C.

Test for glucoses (products of inversion) with alkaline copper or mercury solutions, etc. as described on p. 282.

Hydrolysis of Maltose by Glycase.

Act on 200 c.c. of starch solution with diastase dissolved in water, or fresh malt extract as in experiments on p. 305, till the mixture no longer gives any reaction with iodine.

The solution boiled and filtered till quite clear, contains dextrins and maltose but no glucose : test a portion of the solution by phenyl-hydrazin, and also by acetic acid and cupric acetate, to shew that it contains no glucose. The boiling prevents any further action of the diastase or of enzymes in the malt extract.

To the remainder of the solution add 10 grs. of coarsely ground maize and keep the mixture at 40°—50° C. on the water bath for 5 or 6 hours. Boil and filter till quite clear : the solution contains much glucose, as is easily shewn by the tests mentioned above. A control experiment should be conducted at the same time to shew that the glucose was produced by the hydrolysis of maltose and not from the maize itself.

For this purpose another 10 grs. of coarsely ground maize is digested under similar conditions with about

200 c.c. of water and the mixture is then boiled and filtered. The filtrate is tested for glucose. Some glucose will always be found in this filtrate, but there is generally so much more glucose formed from the solution which contained maltose than in the control solution, that the hydrolysis of maltose in the experiment is clearly demon· strated. If however so much glucose is found in the filtrate from the control solution that it is not obvious that there is more glucose in the filtrate from the maltose solution, then determinations of the glucose must be made in portions of each of the filtrates.

In this case determine

(1)   The original 'reducing power' of the maltose solution.

(2)   The 'reducing power' of the same after the action of the maize.

(The increase of 'reducing power' will be due partly to the formation of glucose by hydrolysis of maltose and partly to sugars formed from the maize itself.)

(3)   The reducing power of the solution obtained in the control experiment.

The amount of glucose in the filtrate from the actual experiment can be calculated from the difference between the values obtained for the ' reducing power' in (2) and (1) by the method explained on p. 285 ; and if the amount of glucose found in (3) be subtracted from this the result gives the amount of glucose formed from maltose in the experiment.

The determinations of 'reducing power' may be made

volumetrically with Fehling's solution, but it is not necessary to remove the dextrins present in these experiments,
as rough results are sufficient to give the chemical evidence
in a perfectly conclusive form.

Decomposition of a glucoside (salicin) by a ferment (synaptase) from another plant.

Add to a 5 p.c. solution of salicin (which gives no
reaction with dilute ferric chloride), some paste of bitter
almonds ground up with water and sand. Digest for a few
hours on the water bath at 40°—50°C., filter clear and test
a portion of the filtrate with dilute ferric chloride, a deep
purple colour is produced (destroyed by acids or alkalies)
due to saligenol, formed by hydrolysis of salicin.

$$C_{13}H_{18}O_7 + H_2O = C_6H_4 . OH . CH_2 . OH + C_6H_{12}O_6.$$
　Salicin　　　　　　　　Saligenol　　　　Glucose.

CHAPTER XVII.

General Experiments.

(1) Determine the increase per day of nitrogenous compounds and cellulose in vigorously growing *Spirogyra*.

Take some healthy *Spirogyra* which has been well washed in distilled water, filter and allow it to drain on a perforated filter plate. Weigh out two equal portions (about 15—20 grs. each). Place one portion (*B*) in ordinary water and expose it to bright sunlight. Dry the other portion (*A*) at 30°, and note its dry weight.

Extract one portion of *A* with 2 p.c. soda and determine the proteids in the extract (nitrogen by Wanklyn's albuminoid ammonia process, see p. 255).

Extract another portion of (*A*) with dilute alcohol and then with 1 p.c. sulphuric acid and treat the residue with ammonia and bromine water (see p. 293): then dry the residue at 100° C. and weigh (= cellulose).

At the end of twenty-four hours take out the whole of *B* and wash thoroughly with distilled water; then collect, dry, and treat in same way as *A*. The increase of proteids and cellulose in B will be the amounts formed per day by the weight of *Spirogyra* taken.

(2)   Shew that formation of starch is influenced by supply of inorganic salts.

Take some young leaves of *Sparganium natans* (or shoots of *Elodea canadensis*) and grow one portion (A) in pure distilled water, and another portion (B) in a culture solution, both being freely exposed to light and air, for several days.

Determine in portions of A the p.c. of starch (if any) and the p.c. of (sulphated) ash.

Do. in portions of B.

(3)   Trace the changes which occur during germination in the reserve materials of an oily seed under different conditions.

Take three equal weights (about 10 grs. each) of air-dried hemp-seeds (A), (B), (C).

Allow B and C to germinate on damp asbestos cloth and when the plumules have reached a length of 2—3 cm., place B in a bell-jar arranged so as to exclude CO_2[1]: place C under similar conditions, but with free access to CO_2: leave B and C exposed as much as possible to light for about a fortnight.   Then kill them by chloroform vapour and dry at $25°$—$30°$.

Note weights (air-dried) of A, B, and C.

Make extracts of aliquot portions of A, B, and C, and determine for each of them, (1) nitrogenous compounds, (2) oils and fats, (3) sugars, (4) starch, (5) cellulose.

[1] See Part I., Fig. 8, p. 30.

APPENDIX I.

MANY of the above experiments give very varying results with different material. As instances of extreme variations we may mention two of our experiences.

On one occasion the shoots of *Onobrychis sativa* which had been kept for the usual time in the dark contained scarcely a trace of amides, and on another no traces of tartaric acid could be detected in beet-root juice, which commonly contains enough of this acid to be easily detected without any special skill in this kind of analysis.

We have therefore added the following notes on our experiences in conducting these experiments with students, and given the numbers obtained for the quantitative results of fairly representative cases.

We have tried as far as possible to arrange the quantitative experiments so that fairly accurate work will clearly demonstrate the principles involved, but in such cases as the estimation of proteids and mixed sugars very careful work is essential.

CHAPTER X. p. 259. *Proteids, etc.*

Qualitative.

The NaOH extract will contain large quantities of proteids insoluble in water and give a copious precipitate when neutralised.

The aqueous extract always contains some proteids soluble in water and may contain a small amount of peptones and albumoses, but these substances are often absent, or present only in traces.

Considerable quantities of amides will be found. Traces of ammonia, nitrates, and nitrites sometimes occur, but commonly these compounds are absent.

Quantitative.

The following values were obtained in a set of experiments with seeds and shoots of *Onobrychis*.

The p.c. are calculated for air-dried material in each case.

	Seeds.	Shoots normal.	Shoots kept in the dark.
Proteids insoluble in water	64·3	17·8	8·0 p.c.
Proteids soluble ,, ,,	2·8	4·1	4·2 p.c.
Peptones and Albumoses	nil	0·7	0·5 p.c.
Amides	trace	0·5	5·9 p.c.
Ammonia Nitrates and Nitrites	nil	nil	nil p.c.

Ammonia. Nitrates, etc. (p. 260).

Experiment No.	1	2	3
	Normal shoots.	Shoots kept in the dark and absence of oxygen.	Shoots from plants watered with ammonium nitrate.
Ammonia	trace	2·8	1·3 p.c.
Nitrates	nil	nil	1·7 p.c.
Nitrites	nil	nil	0·8 p.c.

CHAPTER XI. p. 265. *Oils and Fats.*

Qualitative.

If a portion of the oil from *Lepidium* seeds is qualitatively examined it will be found that the whole of the oil is not saponifiable by aqueous or alcoholic potash. The exact chemical nature of the unsaponifiable substances is uncertain, but they are probably not plastic.

Considerable quantities of glycerin can be detected after saponification of the oil.

Quantitative.

Values obtained with seeds and seedlings of *Lepidium*. The p.c. are calculated for material dried at 100° C.

Experiment No.	1	3	2
	Seeds.	Seeds, germinating radicle protruding a few mm.	Seedlings and remains of seed.
Oils and Fats	30·2	23·7	5·2 p.c.
Glycerin produced by saponifying oil from 100 grs. of material dried at 100°.	3·1 grs.	1·8 grs.	0·5 grs.

CHAPTER XII. p. 275. *Tannins and Glucosides.*

Qualitative.

Experiment No. 1. Willow Bark.

Much tannin is generally present, which is best removed by hide powder; but shaking with lead carbonate will also remove the tannin completely if the process is repeated several times.

Salicin is commonly present in sufficient quantity to be easily detected; and small quantities of glucose are generally found.

Experiment No. 2.

Young and ripe fruits of *Musa*. Much tannin and little sugars may be expected in the young fruits, but in the old fruits much sugars as well as tannins.

Chapter XIII. p. 288.

Qualitative.

Experiment No. 1.

Glucoses, cane-sugar, and maltose are generally all present.

Experiment No. 2.

(*a*)   Very little or no cane-sugar or reducing sugars will be found.

(*b*)   Abundance of glucoses and cane-sugar will be found, but little or no maltose.   Crystals of cane-sugar can easily be obtained by the strontia method.

Quantitative.

Values obtained in a set of experiments with leaves and roots of *Beta vulgaris*.

The p.c. are calculated for material dried at 100° C.

Experiment No. 1.

	Total sugars.		Leaves.	Roots.
Est. by fermentation process and calculated as glucoses			0·2	6·8 p.c.

Experiment No. 2.		(*a*)	(*b*)
	Glucoses	0·5	1·8
	Cane-sugar	nil	nil
	Maltose	0·3	0·7

Chapter XIV. p. 293.

Quantitative.

Values obtained using tubers of potato, leaves of *Acer pseudoplatanus*, and grains and seedlings of wheat.

The p.c. are calculated for material dried at 100° C. in each case.

Experiment No. I.	(*a*)	(*b*)	(*c*)
Starch	57·2	56·9	57·0 p.c.

The results of these three experiments should not differ by more than 1—2 p.c.

Experiment No. II.

Values obtained in three different sets of experiments.

	Leaves killed in the evening.	Leaves killed in the early morning.
Starch	3·4	2·5 p.c.
,,	4·8	1·7 p.c.
,,	4·0	3·2 p.c.

As the figures indicate the result of these experiments varies greatly, and apparently in no regular way, but the greatest difference may be expected when a warm damp night has succeeded a clear bright day.

Experiment No. III.

	Grains.	Seedlings kept in dark for three days.
Starch	59·6	3·0 p.c.

CHAPTER XV. p. 301.   *Organic acids, etc.*

Qualitative.

For ascertaining what organic acids are present, free and combined, beet-root juice may be examined, in which varying quantities of acetic, glycolic, malic, citric, tartaric, oxalic, succinic, and aconitic acids are commonly present.

Quantitative.

Values for old and young petioles of rhubarb.

Experiment No. 1.

Acidity.	Young petioles.	Old petioles.
[Calculated as oxalic acid $H_2C_2O_4$.]	0·6	2·2 p.c.

Values for ripe and unripe apples.

Experiment No. 2.

	Unripe apples.	Ripe apples.
Acidity.		
[Calculated as $H_2C_2O_4$.]	1·2	0·3 p.c.
Sugars.		
Glucoses	0·8	4·6 p.c.
Cane-sugar	nil	0·9 p.c.
Maltose	nil	nil p.c.

The p.c. in experiments 1 and 2 are calculated for fresh tissues.

Using leaves of etiolated and normal potato plants values obtained were

Experiment No. 1 (p. 302).

	Normal leaves.	Etiolated leaves.
Ash (sulphated)	4·7	2·5 p.c.

The p.c. are calculated for leaves dried at 100°C.

Experiment No. 2.

Values for ash from grains and young leaves of wheat.

	Ash from young leaves.	Ash from grains.
P_2O_5	4·0	48·5 p.c.
$\left.\begin{array}{l}K_2O \\ Na_2O\end{array}\right\}$	21·4	7·9 p.c.

The p.c. are calculated for ash—*i.e.* represent weights of P_2O_5 and alkali in 100 grs. of ash of each of the substances not in 100 grs. of each of the substances compared.

Experiment No. 3.

Values for CaC_2O_4 in young and old leaves of *Sempervivum*.

	Young leaves.	Old leaves.
Calcium oxalate (CaC_2O_4)	1·3	4·1 p.c.

The p.c. are calculated for fresh leaves.

CHAPTER XVI. p. 305.   *Unorganised ferments, etc.*

Diastatic Ferments.

Experiment No. 2.

Using the method described in the text (p. 304) an extract of 10 grs. of fresh malt acting for one hour at 50° on soluble starch produced 2·5 grs. of maltose ; whilst a similar extract of 10 grs. of barley grains acting under exactly the same conditions produced only traces of maltose.

Experiment No. 3.

An extract of air-dried leaves of barley compared under exactly similar conditions with the same malt extract as used in experiment No. 2 produced no maltose.

Experiment No. 4.

Adding to the soluble starch solution the finely divided tissues themselves and not extracts which had been filtered clear.

10 grs. air-dried leaf acting for one hour at 50°, produced from soluble starch.	Leaves of *Pisum sativum.*	Leaves of *Trifolium pratense.*
Maltose	0·6 grs.	0·2 grs.

Invertase.   Glycase.   Glucoside ferments.

Experiments No. 1—3.

Positive results should be easily obtained in all these experiments.

In No. 3—where a control experiment is made—we have never found the quantity of sugars produced by digesting 10 grs. of coarsely ground maize with 200 c.c. of water, for 5 or 6 hours at 40°—50°, to exceed 0·2 grs. of 'reducing' sugars.

CHAPTER XVII. p. 312. *General experiments.*

Experiment No. 1.

An experiment under very favourable conditions gave the values

    A (dry weight of *Spirogyra* used)   = 21·5 grs.
    B (dry weight of *Spirogyra* obtained
                after 24 hours) = 24·1 grs.

For twenty-four hours.

                Increase in dry weight = 2·6 grs.

Increase of proteids = ·8 grs.
    (N × 6·3)   } for 21·5 grs. of
Increase of cellulose = 1·3 grs.   *Spirogyra.*

Experiment No. 2.

Using leaves of *Sparganium natans*—sixteen days.

	A in distilled water.	B in a culture solution.
Starch	0·4	6·5 p.c.
Ash	1·1	4·8 p.c.

The p.c. are calculated for material dried at 100° C.

Experiment No. 3.

Values for a set of experiments with hemp seed.
Weights of three portions of original seed.

A.	B.	C.
10·0 grs.	10·0 grs.	10·0 grs.

Weights of A, B and C after experiment.

A.	B.	C.
10·0 grs.	4·2 grs.	11·1 grs.

Analysis of A, B and C after experiment.

	A.	B.	C.
	seeds.	seedlings 14 days without access of CO_2.	normal seedlings.
Nitrogenous compounds $\}$ (N × 6·3)	21·2	14·6	28·4 p.c.
Oils and Fats	37·0	nil	6·2 p.c.
Starch	nil	nil	12·8 p.c.
Cellulose	18·6	63·1	25·5 p.c.
Sugars	nil	traces	traces

The p.c. are calculated in each case for material dried at 30° C.

APPENDIX II.

Reagents required for experiments on Metabolism.

Solids.

Inorganic.

Asbestos.

Lead carbonate.

Magnesia.

Potassium sulphate (acid).

Soda-lime.

Organic.

Aniline acetate.

Brucine.

Cane-sugar.

Dextrose.

Citric acid.

Diphenylamin.

Gelatin.

Hide powder.

Maltose.

Metaphenylenediamin.

Phenylhydrazin.

Phloroglucin.

Salicin.

Starch.

Thymol.

Organic liquids.

Alcohol.   Commercial absolute.

,,   Methylated spirit.

Amyl alcohol.

Benzene.

Chloroform.

Ether.

Ethyl acetate (acetic ether).

Glycerin.

Petroleum ether (completely volatile below 60° C.).

These liquids must leave no residue on evaporation.

[Alcohol.

Alcohol of about 98 p.c. can be obtained from good methylated spirit by distilling two or three times off freshly burnt lime, it should be allowed to stand over the lime for twenty-four hours before each distillation.

The methylated spirit now commonly sold is useless for this purpose but the 'ordinary methylated spirit,' which is still supplied for manufacturing and scientific work, when treated with lime can be used for all purposes where alcohol not stronger than 98 p.c. is required.]

Standard Solutions.

Decinormal Sulphuric acid.
,,   Alkali (baryta).
,,   Potassium permanganate.
,,   Silver nitrate.

Uranium acetate (1 c.c. = ·005 grs. P_2O_5).

Nessler's ⎫ prepared as described
Ammonium chloride ⎬ in Wanklyn's "Water
Alkaline potassium permanganate ⎭ Analysis."

Fehling's.

Sachsse's (alkaline mercuric iodide).

Barfoed's (4 grs. cryst. cupric acetate + 1 gr. acetic acid per 100 c.c.)

Solutions in water.

Acids.

Hydrochloric ⎫
Nitric   ⎬ (concentrated and 10 p.c.)
Sulphuric   ⎭

Acetic (glacial and 10 p.c.)

Oxalic (saturated).
Trichloracetic (10 p.c.)

Alkalis.

Liquor Ammoniæ.
Ammonium carbonate (saturated).
Baryta-water ⎫
Lime-water   ⎬ (saturated).
Strontia-water ⎭
Soda (caustic) (2 p.c., 10 p.c. and 50 p.c.)
Sodium carbonate (saturated).
Copper acetate (10 p.c.)
Copper sulphate (10 p.c.)
Lead acetate (saturated).
Basic lead acetate (prepared by boiling 100 grs. of lead
    acetate and 70 grs. lead oxide (litharge) with 500 c.c.
    of water for half-an-hour and filtering).
Millon's reagent (mercuric nitrate and nitrous acid).
Potassium ferrocyanide (saturated).
    „   ferricyanide (5 p.c.)
    „   nitrite (20 p.c.)
Sodium acetate (20 p.c.)
Sodium phosphotungstate (saturated).
Ferric chloride (10 p.c.)
Bromine-water (saturated).
Litmus (neutral).

Solutions in 98 p.c. alcohol.

Mercuric chloride (saturated).
Thymol (saturated).
Phloroglucin (5 p.c.)
Phenolphthaleïn (3 p.c.)

Special apparatus.

Lunge's Nitrometer.

Soxhlet's Fat-extraction apparatus[1].

Arrangement for continuous agitation.

Material required.

Seeds of Cress (*Lepidium sativum*).

  ,, Sainfoin (*Onobrychis sativa*).

  ,, Hemp (*Cannabis sativa*).

Young and ripe fruits of Apple.

      ,,   Banana (*Musa sapientum*).

Bark of Willow (*Salix viminalis*).

Leaves of Pea (*Pisum sativum*).

  ,, Barley (*Hordeum*).

  ,, Clover (*Trifolium*).

  ,, Tropæolum.

  ,, Beetroot or Mangold-wurzel (*Beta*).

  ,, Sycamore (*Acer pseudoplatanus*).

  ,, (Old and young) of *Sempervivum tectorum*.

  ,,   ,,   Rhubarb (*Rheum*).

  ,, (of normal) Potato.

  ,, (of etiolated) Potato.

Roots of Beetroot or Mangold-wurzel.

Grains of Barley.

  ,, Maize.

  ,, Wheat.

Bitter almonds.

Fresh Malt.

  ,, Yeast.

  ,, *Spirogyra*.

  ,, Shoots of *Elodea canadensis*, or *Sparganium natans*.

---

[1] Schleicher and Schüll have recently introduced paper shells for use with Soxhlet's apparatus in place of the inner tube, which are very satisfactory.

INDEX.

BIOLOGICAL SERIES.

Practical Morbid Anatomy. By H. D. ROLLESTON, M.D., F.R.C.P., Fellow of St John's College, Cambridge, Assistant Physician and Lecturer on Pathology, St George's Hospital, London, and A. A. KANTHACK, M.D., M.R.C.P., Lecturer on Pathology, St Bartholomew's Hospital, London. Crown 8vo. 6s.

British Medical Journal. The editor of the "Cambridge Natural Science Manuals" has been fortunate not only in the selection of the above-named subject but also in securing as authors Drs Rolleston and Kanthack....... This manual can in every sense be most highly recommended, and it should supply what has hitherto been a real want.

The Medical Chronicle. "This handbook is an attempt to supply a practical guide to the post-mortem room," say the authors in their introduction, and any competent reader will acknowledge that they have succeeded in their attempt. They have not only supplied the student with a large amount of reliable information, but have done it in a clear and very readable form.

PHYSICAL SERIES.

Heat and Light. An Elementary Text-book, Theoretical and Practical, for Colleges and Schools. By R. T. GLAZEBROOK, M.A., F.R.S., Assistant Director of the Cavendish Laboratory, Fellow of Trinity College, Cambridge. Crown 8vo. 5s. The two parts are also published separately.

<table>
<tr><td>Heat. 3s.</td><td>Light. 3s.</td></tr>
</table>

Nature. Teachers who require a book on Light, suitable for the Class-room and Laboratory, would do well to adopt Mr Glazebrook's work.

Science and Art. For the practical courses on Heat and Light now forming such a prominent feature in the curriculum of so many of our schools and colleges, these books are admirably suited.

Educational Review. Mr Glazebrook's great practical experience has enabled him to treat the experimental aspect of the subject with unusual power, and it is in this that the great value of the book, as compared with most of the ordinary manuals, consists.

Saturday Review. It is difficult to admire sufficiently the ingenuity and simplicity of many of the experiments without losing sight of the skill and judgment with which they are arranged.

Journal of Education. We have no hesitation in recommending this book to the notice of teachers.

School Guardian. It is no undue praise to say that they are worthy both of their author and of the house by which they are issued.

Teachers' Aid. Text-books of which it would be almost impossible to speak too highly.

Press Opinions.

PHYSICAL SERIES.

Mechanics and Hydrostatics. An Elementary Text-book, Theoretical and Practical, for Colleges and Schools. By R. T. GLAZEBROOK, M.A., F.R.S., Fellow of Trinity College, Cambridge, Assistant Director of the Cavendish Laboratory.

Part I. Dynamics. 4s. Part. II. Statics. 3s.
Part III. Hydrostatics. 3s. [*Nearly ready.*

Educational Review. In detail it is thoroughly sound and scientific. The work is the work of a teacher and a thinker, who has avoided no difficulty that the student ought to face, and has, at the same time, given him all the assistance that he has a right to expect. We hope, in the interests both of experimental and mathematical science, that the scheme of teaching therein described will be widely followed.

Scotsman. While expounding well the theory of the subject, the book is essentially a practical one for use in large classes in schools and colleges. It is simply and clearly written and has a large number of examples, experiments, and illustrative diagrams; and will be welcome to those who have to instruct beginners in the study of Physics.

Educational Times. We are bound to say that the book is full of good matter, clearly expressed, set out in excellent form and good print.

Educational News. We recommend the book to the attention of all students and teachers of this branch of physical science.

Journal of Education. A very good book, which combines the theoretical and practical treatment of Mechanics very happily.

Machinery. It is quite clear that a great deal of care has been taken in the arrangement of this volume, which will be found of great value to students generally whose initial difficulties have been carefully considered and in many cases entirely overcome.

Knowledge. We cordially commend Mr Glazebrook's volumes to the notice of teachers.

Educational Times. The absurdities which infest books on Mechanics, even the very best, in their language involving the term " force " are absolutely avoided.

Glasgow Herald. The student will also find excellent instructions for the working of experiments in the laboratory.

Technical World. The apparatus used is simple and effective and well adapted for classwork.

London: C. J. CLAY AND SONS,
CAMBRIDGE UNIVERSITY PRESS WAREHOUSE,
AVE MARIA LANE.

AND

H. K. LEWIS, 136, GOWER STREET, W.C.
Medical Publisher and Bookseller.